Accessing the E-book edition

Using the VitalSource® ebook

Access to the VitalBook™ ebook accompanying this book is via VitalSource® Bookshelf — an ebook reader which allows you to make and share notes and highlights on your ebooks and search across all of the ebooks that you hold on your VitalSource Bookshelf. You can access the ebook online or offline on your smartphone, tablet or PC/Mac and your notes and highlights will automatically stay in sync no matter where you make them.

1. **Create a VitalSource Bookshelf account at** *https://online.vitalsource.com/user/new* or log into your existing account if you already have one.

2. **Redeem the code provided in the panel below to get online access to the ebook.** Log in to Bookshelf and click the **Account** menu at the top right of the screen. Select **Redeem** and enter the redemption code shown on the scratch-off panel below in the **Code To Redeem** box. Press **Redeem**. Once the code has been redeemed your ebook will download and appear in your library.

DOWNLOAD AND READ OFFLINE

To use your ebook offline, download BookShelf to your PC, Mac, iOS device, Android device or Kindle Fire, and log in to your Bookshelf account to access your ebook:

On your PC/Mac

Go to *http://bookshelf.vitalsource.com/* and follow the instructions to download the free **VitalSource Bookshelf** app to your PC or Mac and log into your Bookshelf account.

On your iPhone/iPod Touch/iPad

Download the free **VitalSource Bookshelf** App available via the iTunes App Store and log into your Bookshelf account. You can find more information at *https://support. vitalsource.com/hc/en-us/categories/200134217-Bookshelf-for-iOS*

On your Android™ smartphone or tablet

Download the free **VitalSource Bookshelf** App available via Google Play and log into your Bookshelf account. You can find more information at *https://support.vitalsource.com/ hc/en-us/categories/200139976-Bookshelf-for-Android-and-Kindle-Fire*

On your Kindle Fire

Download the free **VitalSource Bookshelf** App available from Amazon and log into your Bookshelf account. You can find more information at *https://support.vitalsource.com/ hc/en-us/categories/200139976-Bookshelf-for-Android-and-Kindle-Fire*

N.B. The code in the scratch-off panel can only be used once. When you have created a Bookshelf account and redeemed the code you will be able to access the ebook online or offline on your smartphone, tablet or PC/Mac.

SUPPORT

If you have any questions about downloading Bookshelf, creating your account, or accessing and using your ebook edition, please visit *http://support.vitalsource.com/*

Multi-Stage Flash Desalination

Modeling, Simulation, and Adaptive Control

ENGINEERING SYSTEMS AND SUSTAINABILITY SERIES

Series Editor: **Ganti Prasada Rao**

Co-Editors: **Andrew P. Sage, Heinz Unbehauen, Desineni S. Naidu, Hughes Garnier**

Published Titles

Multi-Stage Flash Desalination: *Modeling, Simulation, and Adaptive Control*
Abraha Woldai

ENGINEERING SYSTEMS AND SUSTAINABILITY SERIES

Series Editor: **Ganti Prasada Rao**

Co-Editors: **Andrew P. Sage, Heinz Unbehauen, Desineni S. Naidu, Hughes Garnier**

Multi-Stage Flash Desalination
Modeling, Simulation, and Adaptive Control

Abraha Woldai

CRC Press
Taylor & Francis Group
Boca Raton London New York

CRC Press is an imprint of the
Taylor & Francis Group, an **informa** business

MATLAB® and Simulink® are trademarks of The MathWorks, Inc. and are used with permission. The MathWorks does not warrant the accuracy of the text or exercises in this book. This book's use or discussion of MATLAB® and Simulink® software or related products does not constitute endorsement or sponsorship by The MathWorks of a particular pedagogical approach or particular use of the MATLAB® and Simulink® software.

CRC Press
Taylor & Francis Group
6000 Broken Sound Parkway NW, Suite 300
Boca Raton, FL 33487-2742

First issued in paperback 2020

ISBN 13: 978-0-367-57565-6 (pbk)
ISBN 13: 978-1-4987-2169-1 (hbk)

Visit the Taylor & Francis Web site at
http://www.taylorandfrancis.com

and the CRC Press Web site at
http://www.crcpress.com

To

Darwish M.K.F. Al Gobaisi

Former Director General, Water and Electricity Department, Abu Dhabi

Contents

Series Foreword

Engineering in the twenty-first century has to take up a different paradigm in the light of global concerns for the sustainability of water, energy, and other life support systems. The conditions for sustainability are on the one hand very simple to comprehend in principle, and on the other hand considerably complex to implement in the real world. Following the broad Brundtland definition (World Commission on Environment and Development [WCED] 1987)—"sustainable development is development that meets the needs of the present generation without compromising the ability of the future generations to meet their own needs"—hundreds of definitions have emerged in different contexts and perspectives. Broadly speaking, books devoted to sustainability science for its concepts, principles, and interpretations and engineering systems in light of sustainability will be considered in this series. Sustainability science is not an easily defined discipline due mainly to its complexity. Problems in its domain are of multiple dimensions and multiple scales. It is the domain of confluence of natural and social sciences.

The books in this series are expected to deal with essential life support systems. Ideally, books presenting sustainable engineering systems are most welcome as are books that promote awareness of the need for sustainability indicating possibilities to achieve the same. Studies on systems for a deeper understanding of their character and potential possibilities for sustainability are also welcome. We proceed with the understanding that industrial processes are sustainable if they operate on renewable energy and materials are conserved by recycling and reuse. In addition, the impact on the land and surroundings must be within certain limits. This is also the so-called industrial ecology paradigm.

This implies the need for better understanding of what, and with what, we build and how we engineer a system. The basic premise is that human-engineered systems for sustainability must fit within the natural systems and operate with resource efficiency and a minimal ecological footprint. This is possible first with a good understanding of the systems we consider and then with the way we deal with them. We need not drastically change our specializations under these circumstances. Our work in the engineering and technology professions has to be conditioned to meet the objective of sustainability. We need to value criteria for "sustainability" in the design and operation of engineering systems just as, for instance, we regard "stability" as important. We will then be able to engineer systems by making suitable adjustments and render them sustainable. Education, research, and development with due importance to sustainability will require a holistic and enhanced understanding of what to make and how to build sustainable systems and operate them sustainably. There will be a clear change from the *business as usual* course.

I feel that the task as the editor for this series is very challenging due to the complexity of its theme, but I am comforted by the assurance of support I have received from my colleagues with whom I have been working for many years. I am grateful to Professors Andrew P. Sage of George Mason University, United States; Heinz Unbehauen of Ruhr University Bochum, Germany; Desineni Naidu of the University of Minnesota, United States; and Hugues Garnier of the University of Lorraine, France, who agreed to advise me in this task as and when necessary. Dr. Gagandeep Singh of CRC Press/Taylor & Francis Group has been enthusiastic and encouraging as I took on this task following some work that I had done earlier for him reviewing books.

This Book

In the sense explained earlier, this book—the first in this series—is a modest step toward a better understanding of multi-stage flash (MSF) desalination plants, which is helpful in designing, building, and optimally operating desalination plants and rendering them resource efficient. Since the early 1990s, I have been associated with this work in a team of which Dr. Woldai has been an active member and I have had the pleasure of learning about MSF desalination and making contributions to it.

While most of the existing books on desalination are either broad in their coverage of desalination processes for completeness or devoted to other specific aspects of desalination, this book is fully devoted to MSF desalination modeling simulation and control in considerable detail. In spite of this specialization, the methods presented in this book are applicable to other processes as well with appropriate modifications. It is hoped that this book will be useful in higher education and research in desalination and water resources. It is also hoped that more and more such efforts will follow the line taken up in this book to pave the way to a well-established system of education and research on sustainable life support systems.

Ganti Prasada Rao
Series Editor

Preface

The work presented in this book may be regarded as a step toward bridging the large gap that prevails between theory and practice in the control of multi-stage flash (MSF) plants and a modest initiative toward a much needed research effort in this direction.

The author's experience in the desalination industry in general and with operating MSF plants in particular over several years has led to the recognition for the need to enhance the state of the art and practice in the modeling, simulation, and control of MSF processes. The work presented here represents a humble contribution to the needed effort in this direction. The author's workplace, distinguished for being the world's largest MSF plant, has provided the appropriate atmosphere and facilities needed for this effort.

After a brief introduction to the water situation in the world, the importance of desalination to meet the needs for water is discussed in Chapter 1. Chapter 2 mentions the various processes of desalination and focuses on the operation and control of MSF plants for large-scale desalination. The MSF process, despite being very simple in principle—as heating, evaporation, and condensation—is energy intensive, and energy is a precious thing that cannot be taken for granted. In order to optimize the performance of MSF plants, a detailed understanding of these plants is necessary in terms of dynamics, operation, and control. The development of a dynamic model based on physical principles has been presented for an 18-stage MSF plant, operating in the United Arab Emirates (UAE). In view of this, Chapter 3 presents the dynamic model of the various elements in an MSF plant in considerable detail. Chapter 4 considers the obtained model and the available measurement data. Some aspects of data reconciliation are presented, and the dynamic model is validated using available data. An analysis of the dynamic model for the purpose of control is presented in Chapter 5. The model is linearized and the control structure is selected for the plant. The resulting dynamic model of the 18-stage plant is found to be of dimension 155. All standard model reduction methods have been attempted, and it was found that the model cannot be reduced to a tractable finite dimensional size. There are many identical stages, each contributing to modes that are not negligible relative to each other. The choice of a nonparametric model and the use of a proportional-integral-derivative (PID) control have been made at this point. In order to efficiently reduce the large-size model to a first-/second-order plus delay model, an optimal method is developed and presented for use in PID control in Chapter 6. Chapter 7 then presents optimal tuning methods for PID control with the optimally reduced models.

Extensive simulations have shown nonlinear behavior in this 18-stage MSF plant model. When linearized at different operating points in the operating region, significant variations in the linearized model parameters were observed. It became clear that controllers tuned at a single-fixed operating point are not satisfactory if the plant has to operate at different conditions.

The prevailing technology is PID control, and the controllers are tuned at *some* operating point and left to remain so. The effects of the controllers that become detuned at other operating conditions are ignored until or unless the consequences are *severe*. The problem is one of controlling a *nonlinear plant* for which sophisticated techniques could be attempted against considerable resistance under the prevailing conditions of practice.

Therefore, a parameter scheduling technique that provides adaptation to the operating conditions is presented in Chapter 8 as a simple practicably acceptable method under the prevailing circumstances. The use of artificial neural networks in this context is also discussed. Chapter 9 includes a discussion on the use of renewable energy sources for desalination and emphasizes the potential of solar energy in the Arab region, which is well known for its arid nature and shortage of water.

For the benefit of the readers, Chapter 10 was added at the end of the main content, and provides descriptions and listings of the programs used in this work. Finally, a Glossary and a global Bibliography are included. The latter includes publications in addition to those referred to and cited in the various chapters for further reading.

Thus, this book is a self-contained treatise on modeling, simulation, and control as applied to practical large-scale MSF desalination plants. I hope that the users of the book will not need to seek help from outside sources within the scope of the subject covered here to capture modeling, simulation, and control of MSF desalination plants. I hope that it will be a useful addition to the literature in the field of MSF desalination.

It is not easy to list all the people who were behind me all these years in this work. Nevertheless, a few very important people will always be remembered. First, I must mention Dr. Darwish Al Gobaisi with whom I have been working for nearly three decades, and a major part of this period was in the Directorate of Power and Desalination Plants, Water and Electricity Department (WED) of Abu Dhabi. Darwish dedicated himself to the development of desalination plants in Abu Dhabi and has seen its impact over the years. His penchant for learning and investigations has led to enormous research and development activities in the WED, which has now been reorganized and renamed ADWEA. Darwish was instrumental in organizing advanced lecture programs and refresher courses for the WED staff on various subjects relevant to power and desalination, including modeling, simulation, and control. Some of us were encouraged to do research. I earned my PhD from the University of Bath, United Kingdom, in which Professors A.T. Johns and R.W. Dunn advised me together with Darwish and Professor Ganti Prasada Rao of the Indian Institute of Technology, Kharagpur, who has been our advisor since the early 1990s. Bushara Makkawi, who has been associated with the WED as the deputy director general, has been very encouraging throughout the years, sharing his vast experience in desalination plants and giving practical, valuable suggestions. Dr. Ahmed Kurdali, Dr. A. Husain, and Dr. K.V. Reddy have also provided helpful suggestions.

I must specially thank Professor Ganti Prasada Rao for providing a great intellectual impetus to our team in general and to me in particular all through the years. He has made research work a joyful experience. He has been strongly urging me to write this book and has provided constant support throughout the endeavor. I thank my family for the patience and understanding not only during the years of work but also while writing this book.

Finally, I must express my deep appreciation to Dr. Gagandeep Singh and Amber Donley of CRC Press/Taylor & Francis Group for the helpful suggestions and cooperation at all stages, from my proposal to the production of the book.

Abraha Woldai

MATLAB® and Simulink® are registered trademarks of The MathWorks, Inc. For product information, please contact:

The MathWorks, Inc.
3 Apple Hill Drive
Natick, MA 01760-2098 USA
Tel: 508-647-7000
Fax: 508-647-7001
E-mail: info@mathworks.com
Web: www.mathworks.com

Series Editor

Ganti Prasada Rao received his BE (Hons) in electrical engineering from Andhra University, India, in 1963, and MTech (control systems engineering) and PhD in electrical engineering in 1965 and 1970, respectively, from the Indian Institute of Technology (IIT), Kharagpur, India. From July 1969 to October 1971, he was with the Department of Electrical Engineering, PSG College of Technology, Coimbatore, India, as an assistant professor. He joined the Department of Electrical Engineering, IIT, Kharagpur, as an assistant professor in October 1971 where he was a professor from May 1978 to June 1997. From May 1978 to August 1980, he was the chairman of the Curriculum Development Cell (electrical engineering) established by the Government of India at IIT, Kharagpur. From October 1975 to July 1976, he was with the Control Systems Centre, University of Manchester Institute of Science and Technology (UMIST), Manchester, England, as a commonwealth postdoctoral research fellow. During 1982–1983, 1985, 1991, 2003, 2004, 2007, and 2009, he visited the Lehrstuhl für Elektrische Steuerung und Regelung, Ruhr-Universität Bochum, Germany, as a research fellow of the Alexander von Humboldt Foundation. He visited the Fraunhofer Institut für Rechnerarchitectkur und Softwaretchnik (FIRST), Berlin, in 2003, 2004, 2007, 2009, and 2011. He was a visiting professor in 2003 at the University Henri Poincare, Nancy, France, and the Royal Society–sponsored visiting professor at Brunel University, United Kingdom, in 2007. During 1992–1996, he was the scientific advisor to the Directorate of Power and Desalination Plants, Water and Electricity Department, Government of Abu Dhabi, and the International Foundation for Water Science and Technology, where he worked in the field of desalination plant control. He is currently a member of the UNESCO-EOLSS Joint Committee.

He has authored/coauthored four books: *Piecewise Constant Orthogonal Functions and Their Applications to Systems and Control*, *Identification of Continuous Dynamical Systems: The Poisson Moment Functional (PMF) Approach* (with D.C. Saha), *Generalised Hybrid Orthogonal Functions and Their Applications in Systems and Control* (with A. Patra) (all three books were published by Springer in 1983, 1983, and 1996, respectively), and *Identification of Continuous Systems* (with H. Unbehauen) published by North Holland in 1987. He is coeditor (with N.K. Sinha) of *Identification of Continuous-Time Systems: Methodology and Computer Implementation* (Kluwer, 1991). He has authored/ coauthored more than 150 research papers.

He is/was on the editorial boards of the *International Journal of Modeling and Simulation, Control Theory and Advanced Technology* (C-TAT), *Systems Science* (Poland), *Systems Analysis Modeling and Simulation* (SAMS), *International Journal of Advances in Systems Science and Applications* (IJASSA), and *The Students' Journal of IETE (India)*. He was guest editor of three special issues: one of *C-TAT* on *Identification and Adaptive Control—Continuous Time Approaches* (Volume 9, No. 1, March 1993), and *The Students' Journal of IETE on Control* (Volumes I and II, 1992–1993), and *System Science, Special Issue Dedicated to Z. Bubnicki* (Volume 33, No. 3, 2007). He organized several invited sessions in IFAC symposia on "Identification and System Parameter Estimation" in 1988, 1991, 1994, and at the World Congress in 1993. He was a member of the International Programme Committee of these symposia during 1988–1997. He was a member of the IFAC Technical Committee on Modelling, Identification and Signal Processing in 1996. He was chairman of the Technical Committee of the 1989 National Systems Conference in India. He is coeditor (with Achim Sydow) of the book series Numerical Insights Series published by CRC Press/Taylor & Francis Group. He is a member of the international advisory boards of the International Institute of General Systems Science (IGSS), *Systems Science Journal* (Poland), and the International Congresses of World Organisation of Systems and Cybernetics (WOSC).

Since 1996, Dr. Rao has been closely associated with the development of the *Encyclopedia of Desalination and Water Resources* (DESWARE) (www.desware. net) and the *Encyclopedia of Life Support Systems* (EOLSS), developed under the auspices of UNESCO (www.eolss.net).

He has received several academic awards, including the IIT Kharagpur Silver Jubilee Research Award 1985, the Systems Society of India Award 1989, the International Desalination Association Best Paper Award 1995, and the Honorary Professorship of the East China University of Science and Technology, Shanghai. He was elected to the fellowship of the IEEE with the citation "For Development of Continuous Time Identification Techniques." The International Foundation for Water Science and Technology has established the "Systems and Information Laboratory" in the Electrical Engineering Department at IIT, Kharagpur, in his honor.

He is listed in several biographical volumes and is a life fellow of the Institution of Engineers (India), life member of the Systems Society of India and Indian Society for Technical Education, fellow of the Institution of Electronics and Telecommunication Engineers (India), life fellow of IEEE (United States), and a fellow of the Indian National Academy of Engineering.

Author

Abraha Woldai received his PhD from the University of Bath (United Kingdom), in 1997. During 1985–1999, he worked in the Water and Electricity Department, Government of Abu Dhabi, in various sections: Solar Desalination Plant, Umm Al Nar Power and Desalination Plant, Process Computers, and Instrumentation and Control. During 1989–1999, he joined the scientific research group in the development of the modeling, simulation, and validation of data for MSF desalination plants to the Directorate of Power and Desalination Plants, Water and Electricity Department, Government of Abu Dhabi, and the International Foundation for Water Science and Technology, where he worked in the field of desalination plant control. Since 1999, Dr. Woldai has been closely associated with the development, from concept to completion, of the *Encyclopedia of Desalination and Water Resources* (DESWARE) and the *Encyclopedia of Life Support Systems* (EOLSS), two major publications of EOLSS Publishers in the United Kingdom, developed under the auspices of UNESCO. Dr. Woldai is presently a member of UNESCO-EOLSS Joint Committee and is on the editorial board of DESWARE.

He has received academic awards from the International Desalination Association (IDA) and Best Paper Award 1995, United Arab Emirates University. He has authored/coauthored several research papers on power and desalination.

1

Introduction

1.1 Water and Life

Life on our planet is supported mainly by water, which exists in three states—solid ice, liquid water, and gaseous vapor—in what is termed as the *hydrosphere* of the earth. In particular, the liquid form that is referred to as *water* is considered to be responsible for life on the planet, supporting its evolution from its beginning and maintaining it throughout. Water transports energy in the form of nutrients to the various parts of a living organism. It has unique characteristics. Living organisms have been able to adapt themselves to varying conditions on our planet. Some plants are able to grow in saline soils and water. Some animals and plants have adapted themselves to very arid conditions existing in some parts of the earth. However, no organism can live without water. Communities of plants and animals thrive in the proximity of water. Human civilizations flourished close to freshwater resources such as rivers and lakes. The special chemical characteristics of water, especially the strong hydrogen bonding between atoms in the water molecule, give this substance some very important and unique properties. For example, water has very strong surface tension; its freezing point and boiling point are very high for its molecular weight; its specific heat is very high; and it anomalously expands upon freezing, forming ice that is less dense than liquid water. Consequently, aquatic life remains safe in extremely cold climates in liquid water under which exist the insulating sheets of the cover.

The quality of water, as the substance of discussion here, widely varies in the different parts of the vast hydrosphere. Water being excellent solvent, natural water is never completely pure. It dissolves a variety of substances during its course of movement in precipitation and passage over and through the ground. It carries a range of dissolved inorganic and organic substances, some picked up from the natural environment, others due to pollution by humans. Also, the biological components in water may include bacteria, viruses, algae, and protozoans, many of which are potential pathogenic organisms. The dissolved, as well as suspended, material profoundly affects aquatic life and the

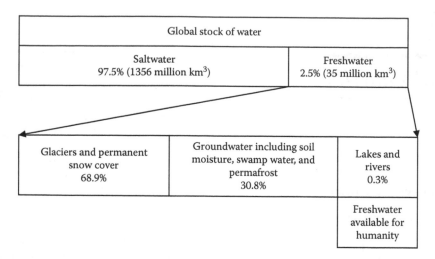

FIGURE 1.1
Approximate distribution of the global stock of water.

usefulness of such water to human life. For human life, freshwater is critically necessary and its scarcity is a major issue for humanity today. On average, seawater contains 35 g of total dissolved solids (TDS) per kg or liter of water, or 35,000 parts per million (ppm). Brackish groundwater generally has a much lower salt content, usually around 2,000–10,000 ppm TDS.

Of the world's water stock that is nearly constant, 97.5% is saltwater from its oceans and only 2.5% is freshwater. Of that 2.5%, approximately 69% is frozen and locked up in glaciers and ice caps, leaving less than 30% in fresh groundwater (swamps account for another 1%) (source: United Nations). Figure 1.1 shows the approximate distribution of water on the planet in different states of quality.

Freshwater resources are renewed by the natural *hydrological cycle* by a process of recycling that is powered by the sun. Annually in this cycle about 577×10^6 km^3 of water is circulated. A great part of this amount is confined to the oceans where water rises as cloud by evaporation and precipitates and returns as rain into the oceans. Only 20% of global precipitation falls over the land surface of the earth and more than half of that quantity evaporates back into the atmosphere. In this process, about 47×10^6 km^3 of water is annually exchanged between the land and the oceans of our planet, thereby renewing the freshwater resources of the earth. Over the land, this amount returns to the oceans by run-off, mainly through rivers. Of the 47×10^6 km^3 of water that drains off the land back into the oceans each year, only about 12,500 km^3 is available for use by humans. Most of the flow is in sparsely populated areas or flows when rivers are swollen after rain, and the volume of water is too great for humans to be harnessed for any use. Much of this water maintains freshwater wetlands and supports ecosystems and biodiversity. The river flow that is intercepted and used by humans is about 2750 km^3/year. There is a further volume of about 3500 km³ freshwater held in storage in reservoirs, from where a variable amount is withdrawn each year.

A vast part of nonsaline water is not accessible for human use as it is locked up in ice sheets and glaciers. Most of the remaining freshwater is groundwater, but the vast part of that is not where humans most need it, and the rate of aquifer recharge is slow relative to withdrawal. We can regard the freshwater available to humanity as that part that is in freshwater lakes and rivers and streams spread over an area of about 105,000 km².

Freshwater resources of the world are not evenly distributed. In some of the arid regions of the earth, they are not adequately touched by the hydrologic cycle, and rainfall and run off are distributed in both space and time with great disparities. Some places on the earth receive enormous quantities of water in this process while other regions receive hardly any. Arid and semi-arid zones of the world, constituting 40% of the earth's landmass, have only 2% of global run-off. The total quantity of water annually available in river flow and underground has dramatically decreased in recent years as a result of human intervention and population growth. There has been a drop of nearly 37% in the water available per person per year over the last quarter of the twentieth century.

Water has always been mankind's most precious resource—there are no substitutes. Water is fundamental to life, health, and development. It is a prerequisite to the realization of all other human rights (UN, 2002). It must be noted that water security, food security, energy security, and economic security are all linked to water. It is one of the most important factors that constrain economic development, social progress, and human survival. Currently, there are more than 100 countries and regions where there is shortage of freshwater, about 1.5 billion people cannot get clean drinking water, 2.0 billion people are living without safe water, and the consumption of water is increasing at a rate of 4%/year. Regions all around the world are currently facing a very serious problem, and people are even dying because of not enough and clean water. The quality and quantity of water are diminishing greatly in third-world countries. Water scarcity has become a major bottleneck that constrains the world's sustainable economic development.

Water security may be defined as "the ability to access sufficient quantities of clean water to maintain minimal standards of food and goods production, sanitation and health." Population growth and economic development are causing a steadily increasing demand for additional supplies of potable water. World water demand, approximately 4200 km³ in 2000, has more than tripled over the past half century and is estimated to be about 30% of the world's total accessible freshwater supply and that fraction may reach 70% by 2025. The current global situation according to the World Health Organization (WHO) estimates that, globally, about one billion people lack access to clean water supplies and that more than two billion lack access to basic sanitation.

If the annual renewable supplies are no more than 2000 m³ per head of population per year, it is considered to be a state of water stress. Renewable resources of 1000 m³/capita/year are regarded as an approximate threshold below which most countries are likely to experience water scarcity on a scale sufficient to impede development and harm human health according to the WHO. Table 1.1 shows estimated annual renewable water supplies for the countries that are already in this condition. The degree of water shortage can be regarded as an index of water stress, which is obtained by subtracting from 1000 the figure for annual per capita water availability. Thus, for example, the projected index figure for the year 2050 for Yemen is 873 (taking the UN's high population projection), whereas that for Morocco is 250.

TABLE 1.1

Projected Water Scarcity for Selected Countries by the Year 2050

| Country | Water Resources (Cubic Meters per Capita) | | |
| | | 2050 UN Population Projections | |
	1990	High	Low
Algeria	690	247	398
Bahrain	184	72	104
Barbados	195	129	197
Burundi	654	160	229
Cape Verde	587	176	252
Djibouti	19	6	8
Egypt		398	644
Israel	461	192	300
Jordan	308	68	90
Kenya	635	141	190
Kuwait	75	38	59
Lebanon		768	1218
Libya		213	276
Malawi	961	236	305
Malta	85	57	88
Morocco		468	750
Oman		163	235
Qatar	103	47	68
Rwanda	902	247	351
Saudi Arabia	284	67	84
Singapore	222	159	221
Somalia	980	223	324
Syria		454	667
Tunisia	540	221	363
United Arab Emirates	293	120	171
Yemen	480	90	127

With the annual renewable resources of less than 1000 m^3/capita/year, a country faces water scarcity.

The values in Table 1.1 are for renewable resources of freshwater, essentially as rain falling on a country or flowing into it from other countries. Augmentation of these resources is not included.

In 1995, 166 million people in 18 countries were below that level and by 2050 clean water availability is estimated to fall below that level for 1.7 billion people in 39 countries. The situation is aggravated by pollution by human activities.

The water crisis is deepened by pollution due to human activities from point sources or nonpoint sources. Point source pollution comes mainly from industrial facilities and municipal wastewater discharges, with or without adequate treatment. Storm drainage and run-off from irrigation, construction sites, and other land disturbances constitute nonpoint sources of water pollution.

Pollutants can be divided into the following seven major groups:

- Pathogens that cause waterborne diseases
- Oxygen-demanding substances that deplete oxygen from water bodies
- Nutrients that support unwanted plant and microbial growth in water bodies
- Heat that reduces the oxygen-holding capacity of water
- Nontoxic chemicals (such as salts) that reduce the beneficial uses of water
- Toxic chemical compounds
- Petroleum compounds (such as oil)

Despite progress in water supply and sanitation technology, and massive investments in efforts to control water-related diseases, the overall global incidence and impact of the diseases would not appear to have decreased. This is due to a number of developments such as

- Increase in the global population of humans and domestic animals
- Frequent and rapid movement of people and goods all over the world
- Deteriorating financial capabilities of many communities and countries
- Climatic changes
- Emerging new pathogens and reemerging pathogens
- Selection for pathogens resistant to water treatment and disinfection processes
- Escalating numbers of people with increased susceptibility to waterborne diseases

In response to the growing water crisis and water scarcity, the following have attracted global attention as indicated by these major international events:

1. World Water Forums in 1997, 2000, 2003, 2006, 2009 (Turkey)
2. UN declared 2003 the International Year of Freshwater and 2005–2015 the UN Decade of Water
3. UN Millennium Summit (2000) and World Summit on Sustainable Development (2002) identified water (and energy) as critical to eradication of poverty and achieving sustainable development

There are mainly two ways of meeting the challenges of scarcity and pollution of freshwater—augmenting freshwater resources by desalination of seawater and brackish water and applying water treatment techniques to recycle used water for reuse. The only practicable methods for large-scale augmentation are extraction of groundwater, which results in lowering of water tables or desalination. Groundwater extraction is essentially a short-term measure that cannot be continued indefinitely, whereas desalination can be continued if energy is available to drive the process.

1.2 Desalination of Seawater

The sea is virtually an inexhaustible source of water in the world and surrounds many arid regions on our planet that have no other viable water resources. With all other resources of water tapped to their limits, and with the gradual growth of arid areas due to environmental imbalances, desalination of seawater is fast becoming a major means of obtaining potable water for the human survival in the long run in many parts of the world. Desalination is a human-engineered process to compensate for the shortage in provision by the natural hydrologic cycle.

Desalination processes can be classified into the following main categories:

1. Thermal processes in which seawater is heated, evaporated, and then condensed to give freshwater.
2. Membrane processes in which seawater or brackish water is passed through semipermeable membranes that block the salt molecules and let freshwater pass to the other side.
3. Electrolytic processes.

The last of these is used to obtain water of very high purity especially for industrial purposes. Thermal and membrane processes are used for large-scale water production.

Among all the seawater desalination processes, the quantity of potable water produced using the multi-stage flash (MSF) process far exceeds that by any other process. In addition to its relatively low cost, the modular structure of an MSF plant is an obvious asset for a facility that must satisfy a variety of production demands. The capability of coupling the MSF plant to a power generation plant as the heat source is an indication that the process will be increasingly exploited in the future to meet the needs of water as well as of power (Al Gobaisi, 2001). However, high-energy consumption is one of the major bottlenecks to limit its promotion and application. Presently, desalination processes are predominantly used in the arid Arab region, where fossil fuels are presently abundant but not renewable. Therefore, it is important to seek methods to reduce the desalination energy consumption and to use energy from renewable resources to ensure water security and thereby sustainability.

1.3 Need for Desalination

The demand for water for domestic and agricultural purposes has increased dramatically since the middle of the twentieth century. To meet this demand, increasing attention has been given to desalting ocean waters and brackish waters in inland seas over the years. There were more than 3500 land-based desalination plants in operation in the early 1990s and a rather large number of smaller plants on board oceangoing ships. Desalination plants are generally located in areas where the population has surpassed the capabilities of the onshore water supply and where presently high-cost desalinated water is affordable. This situation occurs most often in coastal desert areas and densely populated islands. It is less attractive in situations where the cost of pumping desalinated water through pipelines to interior areas adds dramatically to basic desalination costs at ocean side desalination sites.

Today in many Arabian/Persian Gulf countries, a great majority of water supplies are from desalination. Desalination (also called *desalinization* and *desalting*) is the process of reducing the salt content of seawater or brackish waters to an acceptable limit of around 500 ppm (parts per million) TDS, or 0.5 g/L or even lower. It is virtually the only readily available method of obtaining fresh or potable water in arid coastal regions of the world, where sources of freshwater (rivers, streams, or groundwater) are very scarce or totally absent.

Figure 1.2 is an example of a general desalting device that separates saline water into two streams—one with a low concentration of dissolved salts— the freshwater stream—and the other containing the remaining dissolved salts brine streams. It requires energy to operate and can use different technologies for the separation.

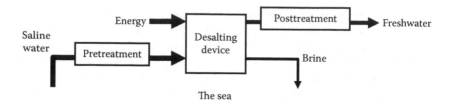

FIGURE 1.2
Simple desalting process.

Desalination technology, originally developed for the treatment of seawater or brackish groundwater, can be modified for the reuse of wastewater for the treatment of industrial liquid streams. It can be used for many applications. The most prevalent use is to produce potable water from saline water for domestic or municipal purposes, but use of desalination and desalination technologies for industrial applications is also growing.

There are many different ways to desalinate saltwater, but the main and most common way for large-scale desalination is called *MSF desalination*, which is a thermal process.

1.4 Basic Principle of Multi-Stage Flash Desalination

MSF desalination is a process in which seawater is heated and allowed to evaporate in the so-called *flash chambers* and condense in the upper parts of these chambers, leaving the salts behind in the water component at the bottom of the chambers as concentrated brine (see Figure 1.3).

The feedwater needs to be heated rapidly such that it quickly reaches the boiling point. The heated seawater is then led into a cascade of chambers at a pressure lower than that of the atmosphere such that boiling at a pressure lower than the atmospheric pressure—called *flashing*—takes place in the flash chambers. In each stage, water vapor is collected and condensed to give freshwater. The resulting clean water is tapped for drinking, irrigation, or industrial usages. Sixty-four percent of desalinated water in the world is produced through MSF distillation. There are plants capable of producing hundreds of millions of cubic meters of water per year or about millions of cubic meters of water per day.

Some facts and features of MSF include the following (Thye, 2010):

- It is a major thermal desalination process—90% of all thermal production and 42% total world desalination production. Thus, it is among the most commonly used desalination technologies.
- It is the most robust of all desalination technologies.

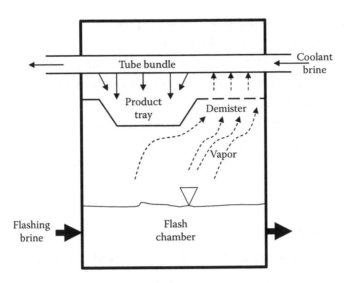

FIGURE 1.3
Single-flash stage.

- It can process water at a very high rate with relatively less maintenance.
- It is capable of very large yields. Plants with design capacities of 600,000–880,000 m³/day are in operation in Saudi Arabia and the UAE.
- It operates using a cascade of chambers, or stages, each with successively lower temperature and pressure, to rapidly vaporize water, which is condensed afterward to form freshwater. The number of stages may be as high as 40.
- MSF operates at top brine temperatures (TBT) of 90°C–120°C—the highest temperature to which the seawater is heated in the brine heater by the low-pressure steam in a cogeneration system. Higher temperatures than this lead to scaling, the precipitation, and formation of hard mineral deposits such as manganese oxides, aluminum hydroxide, and calcium carbonate.
- Its capital and energy costs are quite high, the latter being crucial for sustainability.
- 25%–50% recovery takes place in high-temperature recyclable MSF plant.
- It gives high-quality product water. The total dissolved salts (TDS) of the product of MSF processes are less than 50 mg/L.
- Minimal pretreatment of feedwater required for it.
- Plant process and cost are independent of salinity level.
- Heat energy for MSF can be sourced by combining it with power generation; this is called *cogeneration*.

- However, MSF is an energy-intensive process.
- MSF requires large capital investment.
- MSF has a larger footprint in terms of land and materials.
- Corrosion problems arise if materials of lesser quality are used.
- It has slow startup rates.
- Its maintenance requires shutdown of the entire plant.
- High level of technical knowledge required.
- Its recovery ratio (product rate/seawater feed rate) is relatively low.

1.5 Energy Consumption and Environmental Impacts

Desalination technology, like any other water treatment technology or separation processes, requires the use of energy to produce water. Desalination requires more energy than most other water treatment methods. However, today, developments in desalination technologies are specifically aimed at reducing energy consumption and cost, as well as minimizing environmental impacts.

Desalting costs are reduced by using cogeneration and hybrid processes. Cogeneration or dual-purpose desalination plants are large-scale facilities producing both electric power and desalted seawater.

Electricity and heat are used in desalination and the energy requirements depend on the salinity and temperature of the feedwater, quality of the water produced, and the desalting technology. In addition to electricity requirements, MSF plants use thermal energy to heat feedwater. Thermal distillation methods are particularly suitable for cogeneration as the high-pressure steam that runs electric generators can be recycled in the distillation unit's brine heater. This significantly reduces fuel consumption compared to what is required if separate facilities are built. Cogeneration is very common in the Middle East and North Africa.

1.5.1 Environmental Issues

Increased energy use may cause adverse environmental impacts; the individual and cumulative impacts of energy use and production at a proposed desalination plant will require case-by-case analysis.

The environmental issues of possible concern during construction and operation of desalination plants (mainly large-scale facilities) include impacts that are common to many coastal development projects (e.g., land use and aesthetic impacts), as well as specific impacts associated with the elements of the desalination system and auxiliary infrastructure. In the latter category, these relate to the introduction of highly saline brine and process additives to the marine environment, and to the emission of greenhouse gases and air pollutants due to the energy demand of the desalination process. Other environmental issues of possible concern must be considered, such as entrainment and impingement of marine organisms from the intake of seawater, hazards associated with storage and use of various chemicals, noise, and so on. Most of the potential environmental impacts during the construction and operation of desalination facilities are of a local nature. The potential environmental impacts during operation are mostly continuous, while those associated with construction activities are temporary and mostly reversible.

1.6 Motivation and Aim of This Book

This book is aimed at providing a better understanding in terms of modeling of MSF process in view of its major role in the water desalination industry and its potential for further growth in the future. Advances in process design and materials have so far been satisfactorily incorporated into the current practice. However, in respect of dynamic modeling and control strategies, the state of practice is lagging far behind the reported advances in other fields, for which significant R&D effort is needed. All applicable advances in science and technology should be transferred as quickly as possible to wide industrial applications that benefit human life. Thus, it is for the sake of life support on our planet that desalination in general and MSF processes, in particular, need improved control and immediate efforts in this direction are required.

This book is a step toward bridging the large gap that prevails between theory and practice in the control of MSF plants and represents an initiative toward a much needed research effort in this direction. This book mainly reports on the author's work dealing with dynamic modeling, simulation, identification, and prospects of advanced control for MSF plants.

MSF plants are large and complex. They are also energy and cost intensive, and above all crucial to life support in several regions of the world. Consequently, they are required to meet higher standards of performance, including optimality, cost-effectiveness, reliability, and safety. Needless to say, many of these criteria can be satisfied by improved control, apart from improved plant materials and design. The present-day MSF plant control practice still continues to be by simple PID schemes, which have hardly changed over several decades in the past from their rudimentary form. The conspicuousness of the absence of any significant effort, even in an investigative form, has motivated the author to choose the case of control of MSF plants for study in this book.

The overall objective of this book is to investigate with the aid of simulations, using a model developed from physical considerations and validated by plant measurements, the possibilities for enhancing the quality of the control systems prevailing in MSF plants by optimizing the related PID controllers and rendering them adaptive to the linearized plant dynamic behavior that usually varies within the region of operating conditions due to nonlinearities, or making a robust control design that would take into account the whole set of linear models corresponding to the range of operating conditions.

The success of the project rests on several factors such as a satisfactory dynamical model of the plant, an optimal method of approximating the model to the standard first order plus dead time (FODT) form, a dependable method of simulation and optimization that can handle the original unreduced model, and so on. To ensure these, the book aims to present the following:

1. Development of a detailed dynamic model of an MSF desalination plant from physical considerations and its validation with the aid of measurements on the actual plant.

2. Development of linearized models of the plant in state-space form at chosen operating points.

3. Development of an optimal method of reducing the plant model to FODT form, handled in the continuous-time domain avoiding errors that are known to arise in the existing discrete time treatment.

4. Development of a simulation facility using the plant model in its nonparametric form (impulse/step response) to alleviate difficulties due to high dimensionality of the state space.

5. Optimization of PID control with the unreduced model of the plant.

6. Development of an adaptive control strategy based on parameter scheduling to maintain the optimality of the PID controller over the chosen range of operating conditions.

1.7 Organization of the Book

This chapter has given a brief overview of the situation with desalination using MSF process, and the rest of this book will be devoted to the various aspects of operation, modeling, simulation, and control of MSF processes.

Chapter 2 provides a concise description of the MSF process and its control systems as well as a review of the existing literature for its modeling and control. It is evident from the literature, which shows satisfactory approaches to steady-state modeling, that dynamic modeling needs further attention. The author's work on dynamic modeling, which is presented in the next chapter, is motivated by the current state of affairs.

Chapter 3 presents a mathematical model derived from physical principles to describe the dynamics of an MSF plant. It is simulated on a commercially available flowsheet simulator. The model is tuned with the help of measured data on an 18-stage plant, which is in actual operation.

Chapter 4 presents model validation using extensive measurement data.

Chapter 5 discusses linearization of the present model at certain important operating points and arrives at a pairing scheme by a systematic analysis using RGA. The challenging size of the resulting system model now warrants model reduction that happens to be the subject matter of Chapter 6. Here, several well-known model reduction techniques are shown to fail, making it practically impossible to consider finite dimensional methods of controller design and leaving the ubiquitous PID approaches through non-parametric forms of plant model as the only possibility. Here for the purpose of PID design, an elegant method of optimal model reduction from the original step response obtained by simulating the unreduced model into first- and second-order plus delay forms, as frequently preferred in the process control area, is presented. The reduced models with two levels of approximation at all the chosen operating conditions are presented in the form of transfer function matrices.

Chapter 7 describes optimal PID controller tuning at the chosen operating condition for MSF desalination plant based on several integral performance criteria. Chapter 8 presents two approaches to adaptive control of the plant to pursue optimality under variable operating conditions that influence the plant characteristic due to the inherent nonlinearity of the plant model. One gives a simple parameter scheduling law and the other employs a trained artificial neural network (ANN) that automatically tunes the controller optimally at every point in the operating region. Finally, Chapter 9 summarizes the work and concludes by pointing out directions for future work.

References

Al-Gobaisi, D.M.K.F. (2001), Overview of desalination and water resources, in *Encyclopedia of Desalination and Water Resources (DESWARE)*, EOLSS Publishers, Oxford, UK. www.desware.net.

Thye, J.F. (2010), Desalination: Can it be greenhouse gas free and cost competitive? Report of MEM Masters Project, Yale School of Forestry and Environmental Studies, Haven, CT, 9 May 2010.

2

Operation and Control of Multi-Stage Flash Desalination Plants

2.1 MSF Desalination Process

Multi-stage flash (MSF) desalination is basically an evaporating and condensing process. The heat required for evaporation can be recovered during condensing phase with a subsequent drop in vapor temperature. The MSF evaporator comprises three main sections, namely, the heat recovery stages 1 to NR, heat rejection stages $NR + 1$ to N, and brine heater (Figure 2.1).

Each MSF evaporator stage constitutes a weir box or orifice to control the brine level, demister to intercept brine droplets entrained in vapor and prevent them from being carried over, the condensed vapor (distillate) to the next chamber. The noncondensable gases from the flashed brine move to the vacuum system through the extraction pipes. The brine flows from one stage to the next through the flashing device to give a good flashing pattern. The flashing device has a mechanically adjustable orifice to achieve regulation of the brine level in each stage, and the stages are water sealed from each other to prevent pressure equalization in stages.

The cross section of the chamber of each flash evaporator shown in Figure 2.2 includes the following:

- Flash chamber in which the flashing occurs
- Tube bundle on which the condensation takes place
- Tray to receive the distillate
- The vapor space in between

The basic layout of an MSF process consists of a steam source, a water/steam circuit (brine heater), and an evaporator. The steam is fed to the brine heater to heat the brine. The steam is also used to create vacuum in the evaporator. The plant capacity and performance ratio (PR) dictate the number of stages of the evaporator. Its efficiency mainly depends upon the "flash range," which is the difference between the top brine temperature (TBT) and the discharge temperature. The efficiency is measured in terms of PR, which is the ratio of the

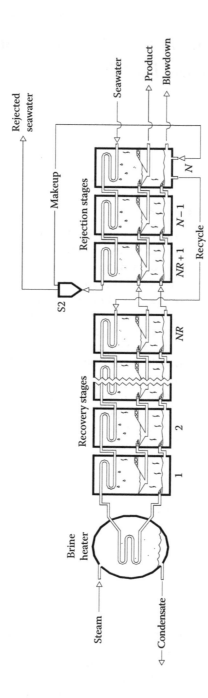

FIGURE 2.1
Schematic flow diagram of an MSF process plant.

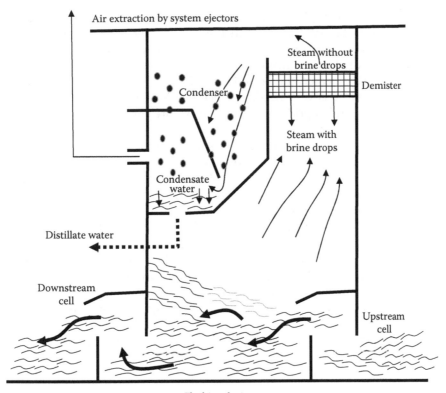

FIGURE 2.2
Cross section of the chamber of a flash evaporator.

product rate to the steam condensation rate. More precisely, it is defined as kilograms of distillate produced per 540 kcal heat supplied by the steam.

The circulating brine is heated by the absorption heat of the distillate and passes to the brine heater where the necessary heating is provided by the steam. This heated brine is flash evaporated in the evaporator cells. The evaporator cells have condensers, through which circulating brine passes, the condensation takes place, thereby producing distillate.

The chlorinated seawater from the seawater supply pump enters the plant and flows through the condenser tubes of the heat reject stages and gains heat through condensation of the vapor from the flashing brine on the outside of the condenser tubes. Most of the cooling water heated in this way is returned back to the sea. A part of the chemically treated seawater from the heat reject section is returned as makeup feed to the deaerator. The required makeup after deaeration is added to the recirculating brine.

The brine is recirculated by the brine recirculating pump through the condensers of the heat recovery stages. Here it absorbs the latent heat of the vapor, thereby raising its temperature and finally flows out from the condenser in the first stage. It then enters the brine heater and is heated by steam to its TBT. In the case of dual-purpose plants, low-pressure steam of extraction or exhaust from the turbine is used as a heat source.

From the brine heater, the heated brine flows into the bottom of the first stage of the heat recovery section. Since it is warmer than the condenser tubes, it flashes and vaporizes and the vapors are condensed by the recirculating brine in the stage.

The brine at lower temperatures flows through the interstage transfer orifices into the next stage. In the subsequent stages, the internal temperature is a few degrees lower than the previous stages, evaporation and condensation are repeated until the brine enters the last stage.

Due to flashing of vapor, the salt content of brine steadily increases from stage to stage. To prevent the increasing salt content in the brine, the concentrated brine is extracted from the last heat reject stage by a blowdown pump and is discharged in to a culvert and an equivalent amount of rejected brine is taken as feed. It should be noted that this makeup feedwater is, in addition to the feedwater, required to replace distillate.

The distillate from each stage is collected at the distillate box in the last stage through the distillate trays and is taken out by the distillate pump, chemically treated for pH and hardness prior to being sent into the water storage tanks.

2.1.1 Interstage Orifice

Interstage flow of the flashing brine occurs through an orifice, which is schematically shown in Figure 2.3, where H_O is the orifice height, L_B is the level of the evaporating brine, P_1 and P_2 are the vapor pressure on either side of the orifice, and y is the vena contracta height (on passage through the orifice).

The interstage brine flowrate in the MSF evaporator depends upon the type and size of the orifice as well as brine levels and pressures on either side of the orifice.

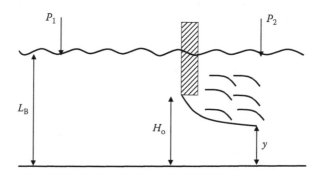

FIGURE 2.3
Interstage flashing flow through an orifice.

The stage pressure, in turn, depends on the operating conditions of the plant, and the levels depend on the hydrodynamics of the stages (Seul and Lee, 1992). High-pressure differences and low-brine flowrates lead to blow-through condition with the accompanying loss in the stage efficiency while the low-pressure differences with high-brine rates result into flooding, leading to loss of evaporation and carryover of salt into the distillate product. Hence, proper type and size of orifice is necessary for a stable operation.

2.2 MSF Desalination Plant Control Systems

About 20 closed-loop controls are used in a large-scale MSF desalination plant (Figure 2.4). Out of these, four main loops, that is, those that control TBT, brine recirculation flow, brine level in the last stage, and seawater flow, are the ones that affect the production rate of the distillate. Among these four loops, the most important control loop in the desalination process that directly affects the production is the control of brine outlet temperature from the brine heater, which is commonly referred to as TBT. The efficiency of the brine heater depends on its design to meet the heat transfer requirements. The overall efficiency and performance of the desalination plant will depend on the type of control structure implemented. The existing brine heater control employed is very simple. This scheme has several shortcomings that are related to the way that disturbances affect the operation of the brine heater.

Conventional controllers with their usual PI, PID action are broadly used for the control of the MSF desalination plants. They function fairly well when the plant is operated at or near the calibrated setpoints. When a disturbance, noise, or system instability occurs, they do not perform satisfactorily because the controller parameters set at the normal operating point are not those needed when the plant disturbance or control loop interactions are encountered.

The following is a brief description of the control systems grouped under the various sections of a typical MSF plant.

2.2.1 Brine Heater Section

2.2.1.1 Top Brine Temperature Control

This is one of the most important control loops in the process used to control the temperature of the brine leaving from the brine heater. The setpoint on the brine maximum temperature is changed, together with brine recycle flow, according to desired distillate production. The recirculated brine that has been heated in the condenser tubes of the heat recovery (gain) section by the condensing vapor from the flashing brine leaves the first stage and enters the brine heater.

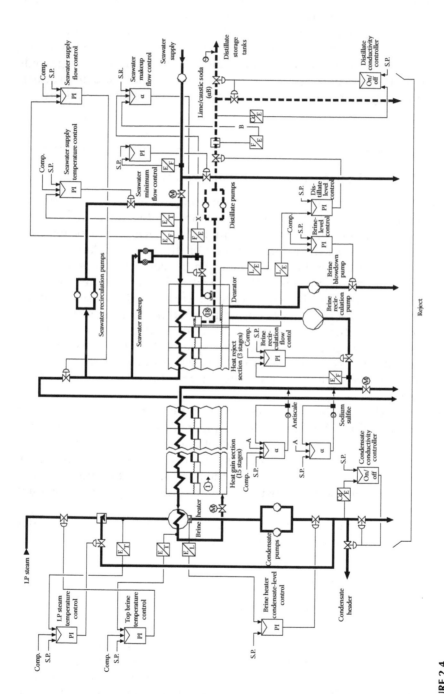

FIGURE 2.4
MSF desalination process. *Note:* S.P., setpoint; Comp., comparator.

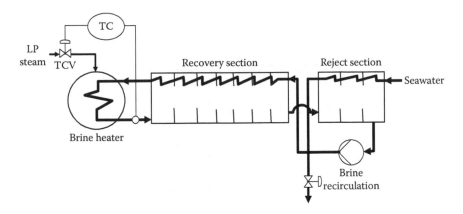

FIGURE 2.5
Top brine temperature control.

The HP/MP steam from the boilers and/or extraction steam from the turbine is reduced to low pressure and is utilized as a heat source to increase the brine temperature in the brine heater to the required TBT. The condensate so formed in the brine heater is either returned to the condensate header or rejected according to its quality. It is important to accurately control the TBT in order to avoid scale formation inside the tubes of the brine heater.

The temperature of the brine at the brine heater outlet is controlled by actuating the LP steam control valve to the heater. The reducing station at the brine heater inlet is designed to cover the required working range. In order to overcome transient conditions and achieve a quick response for the steam pressure in the brine heater, it is normally used as an auxiliary variable in the cascade control mode. The temperature controller has PID control action (Figure 2.5).

2.2.1.2 LP Steam Temperature Control

The purpose of this loop is to control the temperature of the LP heating steam entering the brine heater. The setpoint on the steam temperature is chosen in order to maintain a fixed temperature difference between steam temperature and TBT.

The heating steam in the brine heater is condensed by the recirculating brine flowing through the tubes. Most of the condensate is pumped from the hot well of the brine heater and returned to the steam raising plant.

A part of the condensate is used in the desuperheater to reduce the temperature of heating steam at the inlet of the brine heater. The steam temperature is automatically controlled by increasing/decreasing the spraying water to the desuperheater.

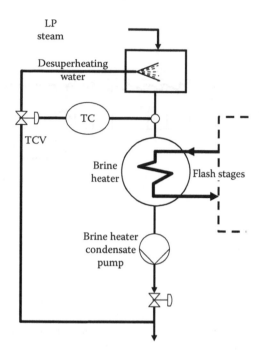

FIGURE 2.6
LP steam temperature control.

In order to avoid the scaling due to high temperatures at the tube walls of the brine heater, a protection circuit is implemented that shuts down the steam supply when temperatures higher than the design temperature are encountered (Figure 2.6).

2.2.1.3 LP Steam Pressure Control

The pressure of the LP steam is controlled by a pressure reduction control valve upstream of the TBT control valve.

2.2.1.4 Brine Heater Condensate-Level Control

The condensate level in the brine heater is controlled by a level control valve on the discharge side of the condensate extraction pump.

2.2.1.5 Brine Heater Condensate Electrical Conductivity (On/Off) Control

The electrical conductivity of the condensate is indicative of its salinity. This on/off control is meant to send the condensate to the boiler or to the reject depending on the salinity of the condensate in accordance with the set limits.

2.2.2 Condenser Section (Recycle Brine Makeup Flows)

2.2.2.1 Brine Recirculation Flow Control

The brine recycle flow control is another essential control of the MSF desalination process because it is directly connected to the distillate production and to the PR. A combination of brine and makeup feedwater is taken by the brine recirculation pump from the last stage and pumped through the condenser tubes of the heat recovery (gain) section. For the stability of plant operation, the brine recirculation flow is controlled and kept constant. Any change in the recirculation flowrate changes the brine levels in all the stages of the distiller and can cause carryover of brine into the distillate. Some means need to be implemented to avoid local boiling of the brine in the heater tubes, which causes scaling in the case of fluctuations in the recirculation flow. The increase in flowrate decreases the flashing efficiency. This is because the brine velocity in the flash chambers increases, thereby reducing the residence time.

The brine recirculation flowrates vary according to summer, winter, high-temperature chemical conditions, and so forth. The distillate output is a direct measure of the brine recirculation and flash range, so these parameters should be measured accurately. The flowrate signal from an electromagnetic flow transmitter is utilized for maintaining the required flowrate in accordance with those given earlier and the required distillate flow by actuating the respective control valve (Figure 2.7).

2.2.2.2 Antiscale Dosing Control

This is a setpoint ratio control. The setpoint is computed from the makeup flowrate and the salinity of the seawater corresponding to the method of scale control. The controlled variable is the flowrate of the chemicals and the manipulated variable is the stroke of the antiscale dosing pump.

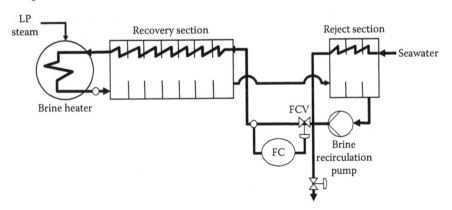

FIGURE 2.7
Brine recirculation flow control.

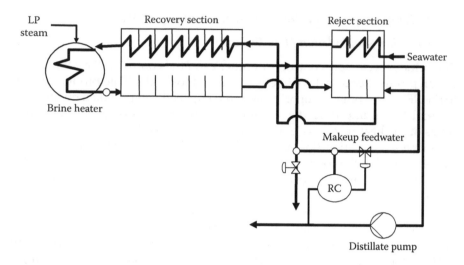

FIGURE 2.8
Makeup flow control.

2.2.2.3 Makeup Flow Control

This is another vital control loop because it is directly linked to the plant distillate production. In order to maintain the brine concentration constant, the seawater makeup flow is varied in direct proportion to the distillate flowrate by a flow ratio controller. The makeup has the scope to reintegrate the produce distillate and discharge brine blowdown. The makeup flowrate is equal to the sum of the distillate and blowdown flowrates (Figure 2.8).

2.2.2.4 Sodium Sulfite Injection Control (Brine Recycle Stream)

Depending on the dissolved oxygen content in the recycling brine, sodium sulfite is injected and controlled in a ratio control loop. The flowrate of sodium sulfite is the controlled variable and the stroke of the dosing pump is the manipulated variable.

2.2.3 Evaporator Section

2.2.3.1 Brine-Level Control (Last Stage)

The brine level in the last stage of the evaporator is maintained at a predetermined setpoint. The level in this chamber has a direct relation to the levels in the preceding stages. The level controller with a PI function maintains the level by actuating the brine blowdown valve.

A part of the concentrated brine in the last flash chamber is discharged into the discharge culvert by the brine blowdown pump. The brine level

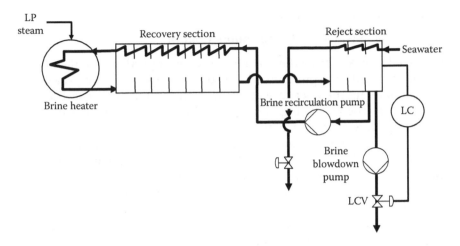

FIGURE 2.9
Brine-level control.

is automatically controlled by means of the level control that increases/decreases the brine blowdown flowrate (Figure 2.9).

2.2.3.2 Distillate-Level Control (Last Stage)

The purpose of this loop is to maintain a preset level of the distillate in the last stage of the evaporator. The setpoint on the level usually depends on the distillate production.

2.2.3.3 Distillate Electrical Conductivity (On/Off) Control

The electrical conductivity of the distillate is a measure of its salinity. This on–off control sends the distillate either to the storage tanks or to the discharge culvert depending on the desired product conductivity.

2.2.3.4 Distillate Blending Injection Control

The distillate is usually blended with seawater to make it potable (drinkable water). This is a ratio control system where the manipulated variable is the flow of the filtered seawater.

2.2.3.5 Distillate pH Control

This is done by injecting caustic soda into the distillate line. This being a highly nonlinear process whose process gain depends significantly on changes of the setpoint or process load, open-loop control is usually applied.

2.2.4 Cooling Water Section

2.2.4.1 Cooling Water Inlet Temperature Control

The heated seawater from the heat reject section is partly recirculated by the seawater recirculating pump in winter to warm up the cooling seawater in order to maintain the inlet seawater temperature to the reject section at the design value. The regulating valve maintains the set value by increasing/ decreasing the recirculation flow.

Since in the winter season the seawater temperature can drop to a low value of about 18°C, the water from the heat rejection section is partly recirculated to maintain the inlet seawater temperature to the heat reject section at the design value. The flowrate of seawater to the heat rejection section is controlled to guarantee the minimum allowable velocity in the condenser tubes. Additionally, a seawater minimum flow line is provided to guarantee minimum flow through the seawater pump (Figure 2.10).

2.2.4.2 Seawater Flow Control (Heat Reject Section)

This is accomplished by manipulating the control valve or by changing the speed of the seawater supply pump. The setpoint is normally determined according to summer and winter modes of operation to maintain the design flow velocities in the tubes (Figure 2.11).

2.2.4.3 Seawater Minimum Flow Control

Seawater flow control is necessary to guarantee a minimum of seawater flow on the suction side of the seawater pump. This protects the seawater supply pump and is normally used during the winter season.

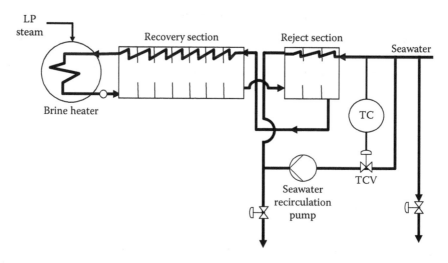

FIGURE 2.10
Cooling water inlet temperature control.

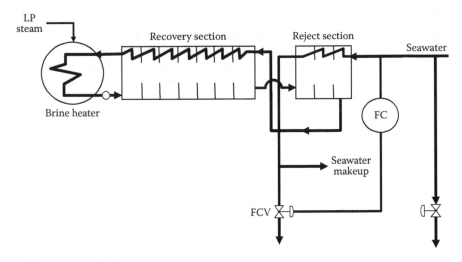

FIGURE 2.11
Seawater flow control (heat reject section).

2.2.5 Ejector and Venting Systems

2.2.5.1 Ejector Condensate-Level Control

This is accomplished by a level control valve at the discharge of the ejector condensate pump.

2.2.5.2 Ejector Condensate Conductivity (On/Off) Control

This on–off system is meant to either send the ejector condensate to line or divert it to the discharge culvert depending on the set limits on electrical conductivity.

2.3 State of the Art in MSF Desalination

In this section, the various developments in MSF technology will be briefly reviewed with a look at the related literature. The purpose of this is to lay the basis, derive motivation and setting appropriate directions for the studies conducted in this work. The author is fully aware of the general process engineering literature that has widely influenced the developments in MSF process engineering, but, in view of its enormity, wishes to avoid detailed references to it. However, a few important publications from the general area of chemical engineering that are relevant to the present context are mentioned, maintaining emphasis essentially on the MSF specialization.

Most of today's MSF desalination plants are based on the design principles suggested by Silver (1957). During the 1960s, several publications appeared on plant design and optimization. The rest of the literature is rich in qualitative information and reports on operational experiences. The interest here being modeling, simulation, and control, this review is organized to cover these aspects separately.

2.3.1 Modeling and Simulation

The objective of modeling any process is to obtain a plausible mathematical description that adequately characterizes the effect of important process parameters on its productivity. Interest in the MSF process models began in the mid-1960s. There are many units in the MSF plant that are quite common in other processes and the modeling methodology for such units is well established (e.g., heat exchangers, pumps, valves, etc.). Models for such units are generally available in the standard textbooks. Therefore, in the present review, only flash stage modeling is considered as it is the basic unit in the MSF plant.

In simulating a plant, all the individual units in the process are separately modeled and coupled together in accordance with the physical interconnections to give the model for the whole plant. The basic approach to modeling is through the mathematical expression of the relationships among the process variables (Barba et al., 1973). These are obtained by considering well-known physical laws, material, energy and momentum balances, and so on, in the form of a set of equations, some of which are algebraic and some differential but generally nonlinear in nature. Some process units with distributed parameters deserve description by partial differential equations. However, for the sake of simplification, lumped approximations are used resulting in ordinary differential equations (ODEs). The analytical and numerical solutions of these equations being very crucial have been the subject of extensive interdisciplinary investigations.

The model should reflect all the important features of the process, it should not be unnecessarily complicated to become unmanageable for computation. The process analysis, based on its mathematical model, therefore, involves three steps: (1) model formulation, (2) development of an algorithm to solve it, and (3) model validation using known information about the process.

Once the model equations are written, it is important to check the consistency of the model, particularly with the large set of complex equations. The number of process variables should be equal to number of equations, so that the degree of freedom is zero in order to obtain a unique solution. If this is not true, the model is either overspecified or underspecified. Another point is to check the consistency of the measurement units in all terms of the model equations.

Mathematical models can be grouped according to several different criteria. Table 2.1 shows a classification of system models according to the four common criteria: applicability of the principle of superposition, dependence

TABLE 2.1

Classification of System Models

Type of Model	Classification Criterion	Type of Model Equation
Nonlinear	Principle of superposition does not apply	Nonlinear differential equation
Linear	Principle of superposition applies	Linear differential equations
Distributed	Dependent variables are functions of spatial coordinates and time	Partial differential equations
Lumped	Dependent variables independent of spatial variables	Ordinary differential equations
Time varying	Model parameters vary in time	Differential equations with time varying parameters
Stationary	Model parameters constant in time	Differential equations with constant parameters
Continuous	Dependent variables defined over continuous range of independent variable	Differential equations
Discrete	Dependent variables defined only for distinct values of independent variables	Time difference equations

on spatial coordinates as well as on time, variability of parameters in time, and continuity of independent variables. Based on these criteria, models of dynamic systems are classified as linear or nonlinear, lumped or distributed, stationary or time varying, continuous or discrete, respectively. Each class of model is also characterized by the type of mathematical equations employed in describing the system. In principle, each industrial process is dynamic in nature; that is, variations or fluctuations with time will always be present in the process parameters with time. If these variations are small enough, they can be ignored and the process can be considered as operating at a steady state; in this instance, time will not be variable and the model will consist of algebraic equations. This model is known as a lumped parameter model. In a steady-state process, however, if spatial variations are to be considered, a distributed parameter model consisting of ordinary or partial differential equations will result depending upon whether a single-spatial coordinate or more than one coordinate are to be accounted for. The steady-state models are mainly used for design purposes as well as for parametric studies of existing plants to evaluate their performance and adjust or optimize operating conditions.

For startup or shutdown conditions of the plant or for control studies, dynamic models are used. In the case of control studies, the process model must be connected to the model of the control system to accomplish simulation of the whole process. The dynamic simulation can be carried out either offline or online. In the first case, there is no connection to the real plant; the input data are fed from a file. In the second case, the input data are directly received from the actual operating plant.

Two types of dynamic models are possible. First, an analytical one (physical), and second, one based on a black-box approach. The analytical model describes the process through physical relations; in addition to the variables considered in the steady-state modeling, the time bound changes in the equipment holdups have to be accounted for. Essentially, the lumped parameter dynamic model consists of an ODE and supporting algebraic equations, both nonlinear; the system is known as a differential-algebraic equation (DAE). In the case of the former type of equations, the initial conditions are either known by experience or developed by steady-state simulation.

In general, the DAE system consists of

$$f(x,\dot{x},y,u,t) = 0 \tag{2.1}$$

$$g(x, y, u, t) = 0 \tag{2.2}$$

where
   t is the time
   $x, y,$ and u are usually referred to as the differential, the algebraic, and the control variables of dimensions n, m, and p, respectively

Similarly, (2.1) is the differential equation and (2.2) is the algebraic equation. In normal dynamic simulation problems, the variation of the controls with time, $u(t)$, is specified by the user, while that of the differential and algebraic variables, $x(t)$ and $y(t)$, is to be calculated.

The other type of dynamic model is based on a black-box approach. A model with unknown parameters is selected according to previous experience or through experiments. Its structure can also be derived from the analytical model by linearizing the equations at the operating point. Then the parameters must be determined by experiments. This step is called *parameter identification*. It may be performed online or offline. For the purpose of control, the parameters are determined during the operation of the plant (online identification). Since the formulation of the analytical model is difficult and complicated, it is common to develop statistical models for control purposes.

The simulation of the true dynamic behavior of a plant requires additionally the dynamic model of the control system. The latter contains the control algorithms of the controller, sensors, transducers, transmitters, and final control elements, which are normally described by linear differential equations (Al Gobaisi et al., 1992).

2.3.2 Solution of the Dynamic Model Equations

Once a dynamic model has been developed, methods are available to solve it numerically. By solve, it means that the transient responses of the dependent variables can be found to some degree of accuracy by numerically integrating the differential equations, given that appropriate initial values for the independent variables have been specified as functions of time.

Over the years, applied mathematicians have developed a large number of numerical integration techniques ranging from the simple (e.g., the Euler method) to the complicated (e.g., the Runge–Kutta method, Gear's method, etc.). All of these techniques represent some compromise between computation effort (computing time) and accuracy. While a dynamic model can always be solved, there may be difficulties in obtaining useful numerical solutions in some cases. With large models, for example, a very short integration interval may be required to obtain an accurate solution; but this may require so much computation time that the solution is impractical to obtain. Collections of software programs for integrating ODEs are available, for example, IMSL package (IMSL, 1988).

Obtaining dynamic solutions of models with large number of equations using standard integration routines may not be straightforward. A number of equation-oriented simulation programs have been developed to assist in this task. The user supplies the set of algebraic and ODEs to the simulation program. The equation-oriented simulator is easy to use, but is expensive in terms of computational burden and effort for the development of the equations for a large process or plant.

An alternative approach is to use so-called modular simulation programs. The approach uses prewritten subroutines to represent an entire process unit. The simulator itself is responsible for all aspects of the solution and includes sophisticated numerical integration procedure. Modular dynamic simulators have been available since the early 1970s, such as DYFLOW (Franks, 1972). SPEEDUP (Perkins and Sargent, 1982) supplies a set of modular programs and physical properties package. Equation-oriented capabilities are directly available to the user. Such simulators are achieving a high degree of acceptance in process engineering and control studies because they allow plant dynamics, alternative control configuration, and resulting operability of a plant to evaluated prior to construction. Other packages for dynamic simulation, like DIVA by Holl et al. (1988), SIMFLOW (Roehm, 1989), and gPROMS, are not available under commercial license.

The first detailed dynamic model for the MSF desalination plant based on basic principles is reported by Glueck and Bradshaw (1970). They divided a flash stage of the MSF plant into four compartments with streams and capacitances interacting materially and thermally among themselves. These are the flashing brine chamber, vapor space, distillate product tray, and the tube bundle transporting the cooling brine. The material and energy balance equations were written for each compartment, with additional salt balance equation for the flashing brine, and noncondensables balance equation for the vapor space. Since no information is available on the mass transfer rates to evaluate vapor generation rates from the flashing brine and the distillate product, their model is based on the following simplifications.

The flashing brine temperature is the sum of the vapor space temperature and the parasitic losses (i.e., boiling point elevation, nonequilibrium allowance, and the demister plus tube bundle losses), the latter are evaluated using available empirical correlations.

The distillate temperature leaving any stage is assumed equal to the condensation temperature.

Little work has been reported to evolve dynamic models of the MSF plants. Specific dynamic simulation programs are developed by Delene and Ball (1971) and Furuki et al. (1985). Rimawi et al. (1989) have solved the dynamic model for a once-through plant by a simultaneous solution.

A detailed survey of these along with a list of software packages is given by Seider et al. (1991). We now look at the literature on modeling some units that are typical of MSF plants.

2.3.3 MSF Plant: Whole Model

The model of a typical MSF plant, complete with all its units including control system components, thus turns out to be a large set of nonlinear algebraic and ODEs for 18-stage plants. The review by Seider et al. (1991) of the various methods of analysis was reportedly motivated by the fact that in practice, designers lag behind the contemporary development of powerful tools for solution and optimization.

In this study, SPEEDUP is used, which is a commercial package marketed by ASPEN Tech to solve the steady-state process and the dynamic process simulation (Pantelides, 1988). Both the steady-state and dynamic simulations were conducted by the present author as reported in Husain et al. (1993).

2.3.3.1 Flash Stage Model

The flash stage is a basic unit in the MSF desalination plant. Superheated brine enters the stage through an orifice from the upstream stage. On its passage through the stage, vapors are generated from it resulting in the reduction in brine temperature since the latent heat of vaporization is supplied by the change in the heat capacity of the brine. The vaporization rate depends on the degree of superheat of the brine, mixing, and residence time. Maximum vaporization occurs at the entrance, accompanied by the reduction in its temperature, due to the maximum driving force and better turbulence. As the brine flows down, the vaporization becomes slow and the change in brine temperature is marginal. No mass transfer correlations are reported in the literature for estimating the vaporization rates. Hence, the only course available is to assume equilibrium between the vapor and liquid phase and correct the estimated rates by using a nonequilibrium allowance, for which some empirical correlations are available (Lior, 1986). In this approach, uniform mixing flashing brine is assumed with respect to temperature and composition though these parameters are distributed in a stage. This assumption is valid since all nonequilibrium effects due to imperfect mixing, bubble growth, and mass transfer at the vapor–liquid interface and lumped into the nonequilibrium allowance.

The distillate product flows concurrent to the flashing brine in a channel, located at the bottom of the cooling tubes. The distillate entering a particular stage will be at a higher temperature than the condensate falling into it. Hence, vapors are generated to cool it down to the condensation temperature. The vaporization normally occurs at the entrance and then any temperature variation of the distillate on its passage through the stage is marginal. It is thus reasonable to assume that the distillate leaving a stage is in equilibrium with the condensing vapor, and the flow pattern is that of the well-mixed stirred tank.

The cooling brine flows in the tubes countercurrent (against) to the flashing brine. Due to the incompressible nature of the liquid, the brine holdup change for any disturbance whatsoever is negligible. Since there is no evaporation in the tubes, there is no variation in the salt concentration. The temperature of the brine changes gradually as it flows from inlet to the outlet way due to condensation of the vapor on the outside surface. Thus, the brine temperature variation is distributed in nature. The outside temperature remains constant since the pure vapor condenses, and the heat transfer driving force is given by the log mean temperature difference. Hence, assuming the cooling brine behaving as well-mixed stirred tank will not lead to major errors in the stage calculations.

Some of the reported features that can be useful in modeling are as follows:

- Corrections for nonequilibrium conditions (Lior, 1986).
- Well-established methods to estimate the physical and thermodynamic properties of water and brine; *Fichtner-Handbook* (Hoemig, 1978).
- Correlation for pressure drop across demister and tube bundle.
- Correlations for calculating inside and outside film heat transfer coefficients.
- The characterization of evaporation from liquid pools as in Gopalakrishna et al. (1987).
- Flash stage hydrodynamic. Some important studies in this direction are by Fujii et al. (1976), Miyatake and Hashimoto (1980), Miyatake et al. (1983a,b, 1992), and Seul and Lee (1990).

2.3.3.2 Brine-Level Model

The estimation of brine level in the stages is a task of considerable complexity despite its apparent simplicity. The major difficulty is due to the complex phenomena occurring in the interstage flow, characterized by bubble rise and growth due to flash evaporation as described by Gopalakrishna and Lior (1987). The orifice coefficient itself is related to the flow conditions.

Hoemig (1978) gave an empirical formula, and Delene and Ball (1971) discussed the complexity of the problem that even today has not been satisfactorily solved. General flow correlations for interstage orifice flow in MSF plants are still a challenge.

2.3.3.3 MSF Plant Control Model

Operation of MSF desalination plants has largely tried to solve control problems by conventional rather than the modern techniques that are now attracting widespread industrial applications and revolutionizing process control as a result of advances in control theory. Control systems designed from mathematical models that are generally imperfect descriptions of real processes. It is widely accepted that conventional PID or PI controllers may not provide satisfactory performance in view of the nonlinearities and uncertainties present in the real world of desalination process. It is essential that control systems operate satisfactorily over a wide range of process conditions.

Presently, most existing MSF desalination plants are controlled by four to six major primary loops: (1) TBT, (2) brine recycle, (3) makeup, (4) brine level, (5) distillate level, and (6) seawater flow, at the drive level without appropriate coordination. This practice was inherited from the 1960s and can be considered to be rather outdated, considering the advances in the area of system theoretical approaches to control system design.

Most of the existing MSF desalination plants are controlled by single-loop controllers. There is indeed a requirement to reexamine the overall control strategy.

The new enhancements and refinements in distiller design accompanied by modern control concepts have opened new ways and means to obtain maximum output with given input. The ever-growing introduction of new chemicals has resulted in operating the new distillers at higher values TBTs. If we are to allow some flexibility in operating a distiller, then an operating envelope can be obtained from the relation of PR versus brine recycle rate, which should be utilized to form points of optimal process operation.

This intricate and interdependent methodology can be implemented in an advanced control system to provide the best possible path with optimum energy and chemical consumption for the desired distillate production rate. Some of the possible improvements toward desalination process control are presently being analyzed, such as coordination of primary control loops at the drive level and also at the higher level (higher level control system of hierarchical functions) (Al Gobaisi et al., 1992). From the point of view of brine heater control, it is certainly appropriate to incorporate flow control in cascade with the TBT. Other possible control strategies would be to consider enthalpy balance for the brine heater control.

Level control should consider both makeup and seawater temperatures as part of this control loop. Another possibility is to couple the brine recycle with makeup in ratio control. This may result in enhancing the level behavior in the last stage. The standard ratio control philosophy should be replaced by algorithmic optimization considering the TBT, brine recycle, enthalpy, and so on.

In the context of a literature survey on MSF plant control, one should consider the vast area of control theory that has had several developments during the last four decades. Other than the classical techniques in the frequency domain with Laplace transforms and transfer functions until the end of 1950s, an era of modern control theory began in the early 1960s giving rise to a wide range of state-space methods. Since the late 1960s, multivariable system theory became the focus of attention. From the early 1980s, the problems of feedback control have been handled using H_∞ and related techniques to yield robust controls in the presence of model uncertainties. Parallel developments took place in the fields of stochastic control, system identification, adaptive control, and so forth. From the mid-1960s, the concept of fuzzy sets of Zadeh and fuzzy logic control has become popular. Techniques of artificial intelligence, expert systems, and neural networks are applied to various problems in control at present. A particular class of optimization techniques known as "model predictive control (MPC)" with variants such as dynamic matrix control, model algorithmic control, internal model control, model-based control, and so on deserve special mention among these developments. There is a big gap between control theory and process control in general. The number of reviews that have appeared in the past discussing this situation is too large to be cited here. Beyond what has been briefly discussed, no formal literature survey will be attempted on "control theory" or "process control" in this chapter in the interest of brevity. The attention is limited in particular to the MSF plant control only.

The practice of MSF desalination process control, process design has remained virtually stagnant and the state of the art of MSF control lags behind contemporary developments in control technology. This is the principal motivation behind the author's present work to attempt at transferring some relevant advances in control technology to MSF control practice as a step toward improved automation of desalination systems that happen to be the lifeline of several regions of the world.

References

Al-Gobaisi, D.M.K.F., A. Hassan, A.S. Barakzai, and M.A. Aziz (1992), Manageable automation system for power and desalination plant, in *Proceedings of DESAL'92*, Al-Ain, United Arab Emirates, pp. 773–814.

Barba, D., G. Linzzo, and G. Tagliferri (1973), Mathematical model for multiflash desalting plant control, in *Proceedings of the Fourth International Symposium on Fresh Water from Sea*, Heidelberg, Germany, Vol. 1, pp. 153–168.

Delene, J.G. and S.J. Ball (1971), A digital computer code for simulating large multistage flash evaporator desalting plant dynamics, Oak Ridge National Laboratory, Oak Ridge, TN, Report #ORNL-TM-2933.

Franks, R.G.E. (1972), *Modeling and Simulation in Chemical Engineering*, Wiley, New York.

Fukuri, A., K. Hamanaka, M. Tatsumoto, and A.S. Inohara (1985), Automatic control system of MSF process (ASCODES), *Desalination*, **55**, 77–89.

Fujii, T., O. Miyatake, T. Tanaka, T. Nakaoka, H. Matsunaga, and N. Sakaguchi (1976), Fundamental experiments on flashing phenomena in a multistage flash evaporator, *Heat Transfer Jpn. Res.*, **5**(1), 84–93.

Glueck, A.R. and W. Bradshaw (1970), A mathematical model for a multistage flash distillation plant, in *Proceedings of the Third International Symposium on Fresh Water from the Sea*, Dubrovnik, Yugoslavia, Vol. 1, pp. 95–108.

Gopalakrishna, S., V.M. Purushothaman, and N. Lior (1987), An experimental study of flash evaporation from liquid pools, *Desalination*, **65**, 139–151.

Hoemig, H.E. (1978), *Fichtner-Handbook on Seawater and Seawater Distillation*, Valkan-Verlag Dr. W. Classen, Essen, Germany.

Holl, P., W. Marquardt, and E.D. Gilles (1988), DIVA—A powerful tool for dynamic process simulation, *Comput. Chem. Eng.*, **12**(5), 421–426.

Husain, A., A. Woldai, A. Adil, A. Kesou, R. Borsani, H. Sultan, and P.B. Deshpande (1993), Modelling and simulation of a multistage flash (MSF) desalination plant, Paper presented at *the IDA and WRPC Conference on Desalination and Water Reuse*, Yokohama, Japan, November 3–6, 1993.

IMSL (1988), *Mathematics and Statistics Library*, IMSL Inc., Houston, TX.

Lior, N. (1986), Formulas for calculating the approach to equilibrium in open channel flash evaporators for saline water, *Desalination*, **60**, 223–249.

Miyatake, O. and T. Hashimoto (1980), Evaporation performance of a compact multistage flash evaporator, *Kagaku Kogaku Ronbun.*, **6**(5), 536–538.

Miyatake, O., T. Hashimoto, and N. Lior (1992), The liquid flow in multistage evaporators, *Int. J. Heat Mass Transfer*, **35**, 3245–3257.

Miyatake, O., T. Hashimoto, and C. Miyata (1983a), Analysis of liquid flow in multistage flash evaporators—Liquid flow pattern and pressure distribution, *Kagaku Kogaku Ronbun.*, **9**(4), 376–382.

Miyatake, O., T. Hashimoto, and C. Miyata (1983b), Analysis of multistage flash evaporation process—Relation between liquid flow pattern and non-equilibrium, *Kagaku Kogaku Ronbun.*, **9**(4), 383–388.

Pantelides, C.C. (1988), SPEEDUP—Recent advances in process simulation, *Comput. Chem. Eng.*, **12**(7), 745–755.

Rimawi, M.A., H.M. Ettouney, and G.S. Aly (1989), Transient model of multistage flash desalination, *Desalination*, **74**, 327–338.

Rohm, H.-J. (1989), SIMFLOW—A combined steady state and dynamic simulator, *CACHI-Symposium, 20th Event of EFCE*, Erlangen, April 1989.

Roehm, H.-J. (1989), *SIMFLOW—A Combined Steady State and Dynamic Simulation*, Dechema-Monographs, 116, VCH Verlagsgesellschaft, pp. 219–228.

Seider, W., D. Dowid, D. Bringle, and S. Widagdo (1991), Nonlinear analysis in process design, *AIChE J.*, **37**(1), 1–38.

Seul, K.W. and S.Y. Lee (1990), Numerical prediction of evaporative behaviour of horizontal stream inside a multistage flash distiller, *Desalination*, **79**, 13–15.

Seul, K.W. and S.Y. Lee (1992), Effect of liquid level on flow behaviours inside a multistage flash evaporator—A numerical prediction, *Desalination*, **85**, 161–177.

Silver, R.S. (September 1957), Method and apparatus for the multistage flash distillation of a liquid, British Patent No. 829, 819.

3

Dynamic Model for an MSF Plant

Dynamic simulation is well recognized as a valuable tool at all stages of plant operation and is suitable for solving problems involving transient behavior, such as studying control strategies, stability problems, process interactions, troubleshooting, reliability, and startup and shutdown conditions (Hiller, 1952; Hamer et al., 1961; Khan, 1986; Heitmann, 1990; Hussian et al., 1992, 1993, 1994; Reddy et al., 1995a; Maniar and Deshpandey, 1996; Peter and Bijan, 2003; Gambier and Essameddin, 2004; Mohamed and Abdulnaser, 2008; Kamali et al., 2009). It is becoming increasingly important, with the emergence of environmental concerns and the need to meet a variety of regulations, to understand at the early stages of design the dynamic behavior of plants and the impact of normal and emergence operations on the environment.

A multi-stage flashing (MSF) process can be modeled mathematically, that is, represented at least approximately, by a set of differential and algebraic equations, whose variables represent particular characteristics of the process. Simulation is the numerical solution of these model equations. Steady-state simulation produces time-independent values of the variables, while dynamic simulation gives the transient solution of the equations.

In this chapter, the dynamic models of the salient units in an MSF plant are derived from physical principles, and the model for the whole plant is built by interconnecting the individual process units as shown in Figure 2.1. The equations include not only the model equations but also all types of correlations to calculate various properties such as densities, thermal conductivities, parameters like heat transfer coefficients, and so on. The inputs will include attributes of all the input streams (flowrate, temperature, pressure, specific enthalpy, etc.) and fixed constructional parameters (tube inside and outside diameters, length, area, etc.). Then, the solution of the dynamic model is an initial value problem for a system of ordinary differential equations (ODEs). At initial time ($t = 0$), all process variables must be known; the steady-state simulation output is a convenient source of such information.

The models were simulated using a commercially available flowsheet simulator (SPEEDUP) package. The system consists of several flash stages of which a few are rejection stages, and the rest are recovery stages, splitter for feedwater, brine recycle and blowdown, brine heater, and so on.

3.1 Literature Review

A dynamic model of an MSF process plant is suitable for solving problems that involve the transient behavior of the plant, such as testing various control strategies, process interactions, troubleshooting, reliability and stability analysis, and startup and shutdown conditions. The true dynamic behavior of a plant can only be simulated by including the model of its control system as well, which is made up of algorithms of the controllers, sensors, transducers, transmitters, and final control elements.

Two types of dynamic models are possible. The first is an analytical model, which is based on physical principles. Hence, time-bound changes in the equipment holdup have to be accounted for. The lumped parameter dynamic process model thus contains ODEs and supporting algebraic equations, both being nonlinear; the system is known as differential-algebraic equations (DAEs). For the ODEs, the initial conditions are either known by experience or calculated by steady-state simulation.

The other type of dynamic model (Al-Shayji and Liu, 2002) is based on a black-box approach, in which a model with unknown parameters is selected according to previous experience or through experiments. Its structure can also be derived from the analytical model by linearizing the model equations at a certain operating point. Then, the unknown parameters must be determined experimentally, which is known as parameter identification and can be performed online or offline. Where the formulation of the phenomenological models is difficult and complicated, statistical models are developed for control purposes. However, the superiority of the former in understanding the true process behavior is unquestionable.

Before reviewing the reported work on MSF dynamic modeling, it is first necessary to discuss *devices* and *connections*, which are conceptually two different types of modeling objects and constitute integral parts of a dynamic model of any chemical process including that of seawater desalination.

3.1.1 Devices and Connections

A device is any delimitable part of a process at a defined hierarchical level of the process decomposition, such as the brine heater in the MSF process or the wall of a single tube in its tube bundle. On the other hand, the connections as denoted by the term itself are entities of the real process, which connect the devices; typical examples are the connecting pipes between devices or solid–fluid phase boundaries in the tube bundle.

The devices and connections occur in an alternating sequence in the process representation. Conceptually, they are distinguished by the roles they play in a real process. A device converts the fluxes of mass, energy, and momentum it receives from its surroundings into a characterizing vector of state variables such as pressure, temperature, concentration, and so on.

In contrast, a connection transforms a driving force, that is, a difference in some potential determined by the known states of two adjacent devices, into a flux. Commensurate with this distinction, only devices possess a nonnegligible volume and thus a holdup for extensive quantities. Hence, in dynamic modeling, only devices and not connections display a holdup. The behavior of the device is usually expressed by differential equations, while that of a connection by a set of algebraic equations mapping forces into fluxes. In addition, coupling information is needed to describe the topology of the structure such as a process flowsheet. Moreover, there are signal transformers motivated by the control system in a chemical process, which do not depict the physicochemical information of a device. Instead, they represent the input/output behavior of the device. Prototypes of signal transformers, for example, are a thermocouple or a proportional integral (PI) controller.

Thus, modeling gives rise to a set of DAEs for each particular object in the process. The set of modeling equations for the entire plant can then be aggregated using the connectivity relations between the different objects. Numerical preprocessing is then applied to transform the set of equations into a form suitable for solution, for example, discretization of partial differential equations (PDEs) by some method of lines. Finally, the degree of freedom and the index of the resulting DAE system have to be examined (Unger and Marquardt, 1991). If high-index problems occur, proper analysis has to be carried out to reduce the index by alternative modeling and/ or choice of design specification (Lefkopolous and Stadtherr, 1993; Unger et al., 1995).

The symbolic preprocessing involves partitioning of the equation system, derivation of the Jacobian matrix, replacing numerically ill-conditioned expressions by well-behaved approximations, and so on. For a state-of-the-art review of several topics as outlined earlier, see Marquardt (1995).

3.1.2 DAE System and Its Index

In general, a DAE system consists of

$$\mathbf{f}(\mathbf{x}, \dot{\mathbf{x}}, \mathbf{y}, \mathbf{u}, t) = \mathbf{0} \qquad (3.1)$$

$$\mathbf{g}(\mathbf{x}, \mathbf{y}, \mathbf{u}, t) = \mathbf{0} \qquad (3.2)$$

in which $\mathbf{x}(t)$ and $\mathbf{y}(t)$ are the *differential* and *algebraic* vectors, respectively, of unknown variables, both functions of time t, while $\mathbf{u}(t)$, also a function of time, is a vector of known variables. Normally, Equation 3.1 arises from dynamic material, energy, and momentum balances. On the other hand, much faster processes like thermodynamic equilibria yield type 2 algebraic equations; all auxiliary equations are also of that type. If, for given values of $\mathbf{x}$, Equation 3.2 is solvable for $\mathbf{y}$, then the DAE system can be converted into

ODE form. However, such transformation is not convenient when Equation 3.2 is nonlinear and has to be solved numerically in each step of integration.

For Equations 3.1 and 3.2, the consistent initial condition is a set of vectors, $[\mathbf{x}(0), \mathbf{y}(0), \dot{\mathbf{x}}(0)]$, which must satisfy both the equations. Although this requirement may be sufficient for many DAE systems, there are some DAEs in which differentials of some of the equations in the system with respect to time enhance the consistency requirement of the set of initial conditions. This is illustrated by the following trivial example from Pantelides et al. (1988).

3.1.3 Example: Linear DAE System (A)

$$\dot{x}_1 = x_1 + x_2 + y \tag{3.3}$$

$$\dot{x}_2 = x_1 - x_2 - y \tag{3.4}$$

$$x_1 + 2x_2 - y = 0 \tag{3.5}$$

In this system, Equation 3.5 can be used to eliminate the algebraic variable y from Equations 3.3 and 3.4, thus converting them into ODEs in x_1 and x_2. As such, Equations 3.3 through 3.5 are a set of three equations in five unknowns $x_1(t)$, $x_2(t)$, $y(t)$, $\dot{x}_1(t)$, and $\dot{x}_2(t)$.

Thus, for initial values at $t = 0$, two can be arbitrarily fixed and the remaining three are obtained by solving the equations. In this system, note that arbitrarily specifiable conditions are equal to the number of differential equations in the system.

Consider linear DAE system (B) in which the differential equations are the same as Equations 3.3 and 3.4; however, Equation 3.5 is replaced by the following equation:

$$x_1 + 2x_2 = 0 \tag{3.6}$$

Note that the DAE system (B) comprising Equations 3.3, 3.4, and 3.6 cannot be converted into ODEs, since it is not possible to eliminate algebraic variable y. Moreover, arbitrary values $x_1(0)$ and $x_2(0)$ cannot be chosen, since the two are now related through Equation 3.6. In fact, the differential of Equation 3.6 with respect to t, that is,

$$\dot{x}_1 + 2\dot{x}_2 = 0 \tag{3.7}$$

must also be satisfied by any consistent set of initial conditions. Hence, Equations 3.3, 3.4, 3.6, and 3.7 form a set of four independent equations in five unknowns: that is, $x_1(t)$, $x_2(t)$, $y(t)$, $\dot{x}_1(t)$, and $\dot{x}_2(t)$.

So only one of these variables can be arbitrarily fixed despite the fact that as earlier, there are two differential equations in the original set of Equations 3.3, 3.4, and 3.6. Thus, in spite of their apparent similarity, DAE system (A), comprising Equations 3.3 through 3.5 and system (B), comprising (3.3), (3.4), and (3.6), are qualitatively different. This difference in DAE system is expressed by their *index*. The *index* is defined as the minimum number of differentiations with respect to time that should be done to convert the DAE system into a set of ODEs. According to this definition, any ODE system has an index of zero. DAE system (A) has an index of one, since a single differentiation of (3.5) gives ODEs. But DAE system (B) has an index of two. The first differentiation leads to Equation 3.7; then using Equations 3.3 and 3.4, $\dot{x}_1$ and $\dot{x}_2$ are eliminated to give

$$3x_1 - x_2 - y = 0 \tag{3.8}$$

A second differentiation applied to Equation 3.8 yields

$$\dot{y} = 3\dot{x}_1 - \dot{x}_2 \tag{3.9}$$

In this way, two differentiations convert the DAE system (B) into ODEs Equations 3.3, 3.4, and 3.9. The initialization of most index one problems is quite similar to that of the ODEs, but some index one and definitely those with higher index problems face difficulties as demonstrated in the case of DAE system (B). Hence, to reduce the index, it becomes necessary in such cases to consider differentiation of some of the equations in the original system, with or without subsequent algebraic manipulations. However, as pointed out by Lefkopolous and Stadtherr (1993), any differentiation of the original system presents several problems including loss of information. It is, therefore, always preferable to deal with a DAE system in its original form having an index of one. An algorithm suggested by the same authors helps to select from among different sets of independent equations and variables in index one problem formulation. If this algorithm fails, alternative equations and modeling assumptions should be considered to find one such desired formulation.

3.1.4 Dynamic Model Solution

As in the steady-state model, the number of equations in the dynamic simulation should be equal to the number of variables plus the number of inputs in order to obtain a unique solution. Here, the equations include not only the model equations but also all types of correlations to calculate various properties such as densities, thermal conductivities, parameters like heat transfer coefficients, and so on. The inputs include attributes of all the input streams (flowrate, temperature, pressure, specific enthalpy, etc.) and fixed constructional parameters (tube inside and outside diameters, lengths, areas, etc.). The dynamic model is then constituted by an initial value problem for a system of ODEs. At initial time ($t = 0$), all process variables must be known; the steady-state simulation output is a convenient source of such information.

The main problem in the dynamic simulation of an industrial-scale process is the solution of a large system of DAEs. For this, not only efficient numerical methods but order reduction methods are also necessary, but one has to contend with the loss of accuracy associated with the latter.

A system of DAEs can be integrated using a standard initial value integrator such as the Runge–Kutta (RK) method or Gear's method. However, the solution of $y(t)$ requires that the Jacobian matrix of Equation 3.2 should not be singular. If the Jacobian is invertible, the index of DAEs is one. Alternatively, Equation 3.2 can be differentiated to convert it into an ODE, so that the total system of ODEs is solved by using a standard routine. Consistent initial conditions are to be given for the solution of the resulting ODEs. DAEs with an index of more than one are problematic in providing consistent initial values. In some cases, it is possible to transform a higher index system into one having an index equal to one. The index can also be lowered by replacing ODEs by a set of nonlinear algebraic equations. It may also be possible in some other cases to avoid the higher index with a proper choice of design specifications and process modeling (Lefkopolous and Stadtherr, 1993).

The solution of dynamic models, as in the case of the steady state, can be obtained stage by stage or simultaneously. The general-purpose dynamic simulators like SPEEDUP (Aspen Tech, 1991), DIVA (Holl et al., 1988; Kroner et al., 1990), and QUASILIN (Smith and Morton, 1988) use methods for the simultaneous solution of model equations, whereas DYNAMIC and FLOWPACK II of ICI solve stage by stage. A knowledge-based, flowsheet-oriented, user interface for DIVA has been discussed by Bar and Zeitz (1990). It is concerned with the structuring of the factual knowledge of the chemical engineering modeling domain.

3.1.5 MSF Modeling

The first attempt in this direction was by Glueck and Bradshaw (1970), who divided a flash stage into four compartments, with streams and capacitances interacting materially and thermally. However, no simulation results were provided. Moreover, their model is over specified because of a differential energy balance combining vapor space and distillate in the flash stage.

Delene and Ball (1971) also considered four compartments in a flash stage. For a better representation of the cooling brine holdup inside the tubes, they were divided into two well-mixed tanks. The noncondensables in the vapor were not accounted for. To calculate evaporation rates and interstage flow, plant-specific correlations were used. Ulrich (1977) applied this model in simulating a test plant containing six flash stages and found reasonably good agreement between the measured and simulated results on disturbing the steam temperature and brine recycle rates. But significant deviations were noted for disturbances in the cooling water rate, for which no explanation was given.

Fukuri et al. (1985) and Rimawi et al. (1989) solved the dynamic model for a once-through plant with a simultaneous solution. The latter observed the trend of various variables for 15 s.

The author with other research groups (1992, 1993, and 1994) developed a dynamic model considering the flashing and cooling brine dynamics and later the distillate dynamics. The models were solved using the SPEEDUP flowsheeting package. For the reduction of steam flowrate to the brine heater by 26%, the simulation results were compared with the available plant results for the same reduction in the steam flow, noting good agreement between the two. The open-loop response of the top brine temperature (TBT) for a step change in the steam flowrate was compared with the actual plant test data. Both the simulation and the plant test results agreed to a sufficient degree qualitatively.

Since seawater or brine is a solution of electrolytes, Marquardt (1996c) attempted steady-state and dynamic models of the MSF process in terms of electrolyte thermodynamics characterized by simultaneous physical and chemical equilibria. Two different types of dynamic models are proposed, namely, a white-box thermodynamic model consisting of balance equations for each atomic species and a black-box thermodynamic model in which equations are formulated in terms of apparent component concentrations. The latter was implemented in the SPEEDUP using a steady-state flash routine from Aspen Plus capable of electrolyte calculations.

3.1.6 Holdup and Interstage Orifice Flow

The hydrodynamics of a flash stage are influenced by the interstage flow arrangement and stage internals provided. Sluice gate orifices are the most used orifices between the stages for flow regulation. Many types of stage internals are used to achieve better stability, desired turbulence, and maintain proper brine levels for different operating loads and for specific orifice openings. A box or weir type arrangement at the exit of the sluice gate or a kick-plate away from the sluice gate is the most commonly used internal.

Submerged or nonsubmerged types of brine flow regimes are desirable in the flash stage of an MSF desalination plant. The type of flow regime depends upon the orifice size and the type of (see Figure 3.1) operating conditions, such as temperature and brine-level differences in the adjacent stages and the brine flowrate. Both types of flow regimes commonly occur in the stages provided with kick-plates, but only the submerged flow occurs in the other types of arrangements. In the submerged flow, the average brine level at the inlet and the outlet in a stage gives a reasonable estimate of the brine holdup. On the other hand, in the nonsubmerged flow, it is necessary to estimate the length and height of the hydraulic jump and the toe of the jump to evaluate the stage holdup accurately. Though holdup is important in

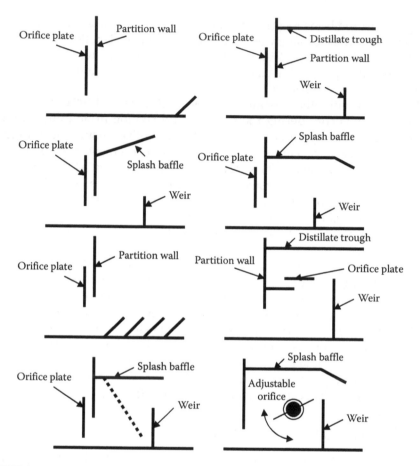

FIGURE 3.1
Examples of interstage brine transfer configurations.

the estimation of the nonequilibrium allowance, no work has been reported to estimate it accurately in the MSF stage.

In an open-channel flow, hydraulic jump occurs when the supercritical flow changes to subcritical due to frictional resistance of the floor or an obstruction placed on the floor. The jump may be submerged or nonsubmerged depending on the tail water depth, that is, the depth at the exit of the stage. In fact, these jumps are intended to dissipate energy and to obtain better turbulence in the flow stream. Energy dissipation is less in the submerged hydraulic jump than in the nonsubmerged one. While it is easy to predict the type of flow, whether submerged or nonsubmerged, by making a momentum balance, it is difficult to predict the shape and toe of the hydraulic jump based on the basic principles. Chow (1959) reported experimental data on these aspects, from which an empirical correlation can be developed to predict these properties.

The interstage brine flowrate in the MSF evaporator depends upon the type and size of the orifice as well as brine levels and pressures on either side of the orifice. The stage pressure, in turn, depends on the operating conditions of the plant and the levels depend on the hydrodynamics of the stages. High pressure differences and low brine flowrates lead to blowthrough conditions with the accompanying loss in stage efficiency, while the low pressure differences with high brine rates result in flooding, leading to loss of evaporation and carryover of salt into the distillate product. Hence, proper type and size of the orifice is necessary for a stable operation.

Ball (1986) discussed the application of Henry's model for the submerged flow, which considers conservation of energy across the orifice and conservation of momentum between the downstream side of the orifice and a point downstream of the submerged hydraulic jump. The limitations of this model were pointed out in application to flow-box experimental data collected at Oak Ridge National Laboratory (ORNL) (Dearth et al., 1970) due to the following reasons:

- Nonavailability of an effective value for the contraction coefficient in the complete range of operation
- Sensitivity of flowrate to upstream and downstream liquid levels

These experimental data were fitted by a quadratic type of correlation for flow as a function of three dimensionless variables, namely, ratios of upstream and downstream liquid levels to orifice height and pressure difference between stages to orifice height.

For the weir type of orifice with submerged flow, Kishi et al. (1985) developed a model using the basic principles of an incompressible fluid flow. The model consists of the two equations for the orifice as used in Henry's model plus the conventional weir equation; all the equations were solved simultaneously. The model was tested against the data from a prototype and a commercial plant, noting excellent agreement between the model output and operating data. They further observed that a wider range of operation can be performed by changing the weir height instead of the gate opening.

Furthermore, for the box-type orifice, Kishi et al. (1987) suggested a flow model considering the analogy between the two-phase flow and the compressible flow. Bernoulli's equation for the compressible flow was applied between various points, namely, orifice inlet to box inlet, box inlet to box orifice, and box orifice to box outlet, solving all the equations simultaneously to obtain the interstage flow. They considered the critical flow characteristics of two-phase flashing flows by calculating the Mach number. Thus, the levels estimated for prototype and commercial plants were in good agreement with those measured.

Hillal and Marwan (1985) proposed a two-phase flow model by placing a baffle at the downstream of the conventional sluice gate. This type of arrangement, which is similar to the box-type orifice, stabilizes the interstage flow.

A slip flow model consisting of momentum balance equations was used for both phases separately. Such a two-phase model is more realistic since similar conditions prevail in the sluice gate. However, their model was not validated against any experimental or plant data.

Large amounts of data are reported in the hydraulics literature for flow through sluice gates, but nothing specific for the MSF process stages. Gerd Posch (1977) and El Hisham (1981) reported studies on specific types of orifices of different sizes in the MSF flash stages. The effect of various parameters on the discharge coefficient was studied.

3.2 Brine Heater Model

The brine heater is one of the principal equipment items in the MSF plant. It heats the incoming recycle stream to the maximum temperature called the TBT. Models incorporating different details can be derived and implemented for this process unit.

Figure 3.2 shows a typical scheme for the steam flow containing a desuperheater in which condensate is injected, a condensate-level controller,

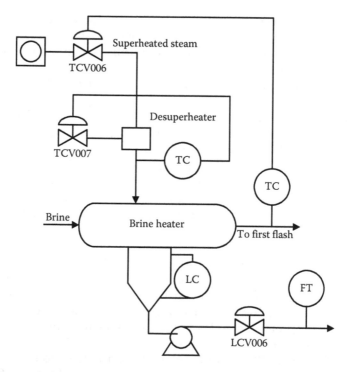

FIGURE 3.2
Steam flow scheme.

and a few valves. All these components have to be included in the brine heater assembly model. The following assumptions are in order:

1. Changes in the potential and kinetic energies are negligible.
2. The condensate channel and water boxes are perfectly mixed.
3. Compression work at the liquid surface is negligible in comparison to thermal energy.
4. The density and specific heat capacity of the brine are constant over the operating range; if necessary, rigorous thermodynamic property correlations can be used by removing this assumption.
5. The vapor phase is taken as lumped quasi-stationary; the time constants of the vapor phase are much smaller than those of the ancillaries.

Now, the following alternatives can be considered for brine heater modeling.

1. *Liquid in tube*: Lumped or distributed
2. *Tube wall capacitance*: Ignored or distributed
3. *Shell capacitance*: Ignored or lumped
4. *Water boxes*: Adiabatic or nonadiabatic
5. *Water box capacitance (input/output)*: Ignored or lumped
6. *Heat flux*: Logarithmic mean temperature difference or local temperature differences

3.2.1 Model Version 1

In this simplest version, water boxes are taken as adiabatic, liquid in tubes as lumped, and the logarithmic mean temperature difference is used for heat flux calculations. The capacitances of tube wall, shell, and water boxes (headers) are neglected.

Water box (input and output side):

$$M_{WB1}C_{B,WB1}\frac{dT_{B,WB1}}{dt} = BC_{B,WB1}\left(T_{B,in} - T_{B,WB1}\right) \tag{3.10}$$

$$M_{WB2}C_{B,WB2}\frac{dT_{B,WB2}}{dt} = BC_{B,WB2}\left(T_{B,T} - T_{B,WB2}\right) \tag{3.11}$$

Brine in tubes:

$$M_T C_{B,T}\frac{d\bar{T}_{B,T}}{dt} = BC_{B,T}\left(T_{B,WB1} - T_{B,T}\right) + Q \tag{3.12}$$

$$\bar{T}_{B,T} = \frac{T_{B,in} + T_{B,T}}{2} \tag{3.13}$$

Vapor phase:

$$Q = W_s(H_s - H_c) \tag{3.14}$$

$$T_s = T_s(P_v, H_s) \tag{3.15}$$

$$H_c = H_c(T_s, P_v) \tag{3.16}$$

Condensate channel:

$$\frac{dM_c}{dt} = W_s - W_c \tag{3.17}$$

$$M_c C_c \frac{dT_c}{dt} = W_s C_c \left(T_s - T^*\right) - W_c C_c \left(T_c - T^*\right) \tag{3.18}$$

Heat flow:

$$Q = U_H A_H \Delta T \tag{3.19}$$

$$\Delta T = \frac{T_{B,T} - T_{B,in}}{\ln\left(\dfrac{T_s - T_{B,in}}{T_s - T_{B,T}}\right)} \tag{3.20}$$

The following are constant parameters of the model equations:

M_{WB1}, W_{WB2}, M_T	Mass holdups of brine in water boxes and tubes
$C_{B,WB1}, C_{B,WB2}, C_{B,T}, C_c$	Specific heat capacities of brine in water boxes and tubes, and of condensate
A_H	Total heat exchange area of the heater
T^*	Standard temperature for enthalpy evaluation

The following process quantities have to be specified as inputs:

$B, T_{B,in}$	Recycle brine flowrate to heat exchanger and its input temperature
P_v, H_s	Thermodynamic properties of steam
W_c	Condensate flowrate from the condensate channel
U_H	Overall mean heat transfer coefficient

With these constant parameters and inputs, the solution of model Equations 3.10 through 3.20 would provide a unique solution. Moreover, the DAEs in the model constitute an index one set; any five independent variables can be chosen for the feasible initial conditions.

Disturbances to which the brine heater can be possibly exposed are the input flow of the brine from the heat recovery section and the thermodynamic state of the steam supplied. Hence, the following two cases are studied using model version 1 in Deutsche Babcock (1994).

1. A 5% step increase in the input brine temperature
2. A 22% step change in the steam pressure with a constant degree of superheating of the low-pressure steam entering the brine heater

Figure 3.3 shows the results of simulation for a step increase in the input brine temperature (Seifert and Genthner, 1991). The displayed brine outlet temperature from the second water box (marked as brine-heater water box (2) T_aus), representing the TBT, shows an initial reverse response. This can be expected, since the version 1 model is not capable of reproducing the dead time resulting from the distributed character of the liquid flow inside the tubes. In this model, the inner energy is calculated by averaging the energy at both ends of the tube, while the energy at the tube entrance is subjected to a first-order change due to the capacitance of water box 1. As a result, the TBT shows an initial decrease, because the brine with the higher energy content has not yet replaced the liquid present there. In this particular instance, the situation can be improved by assuming the liquid in the tubes to be perfectly mixed; however, a more sluggish response of the liquid temperature is to be expected.

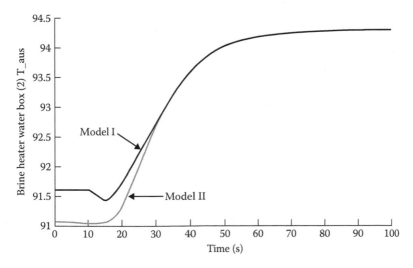

FIGURE 3.3
Step response to a 5% increase in input brine temperature.

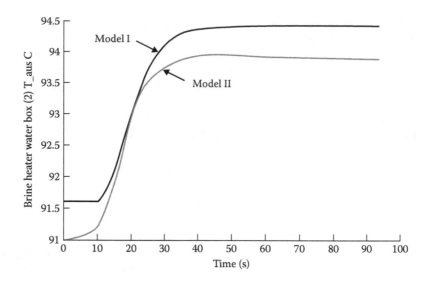

FIGURE 3.4
Step response of liquid temperatures 1 to a change in steam pressure.

Figure 3.4 shows the response of liquid temperatures to a step change in the steam pressure.

3.2.2 Model Version 2

This model differs from that of version 1 in the following respects:

1. The liquid in the tubes is taken as distributed instead of lumped; axial dispersion in the liquid and in the wall is neglected.
2. Tube wall capacitance is explicitly accounted for and considered as distributed.
3. In heat flux calculations, local temperature differences are used.

The rest of the conditions remain the same as in version 1. Equations 3.10 and 3.14 through 3.18 of version 1 remain the same.

Brine in tubes:

$$\rho_B C_{B,T} \frac{\partial T_{B,T}}{\partial t} = \rho_B u_B C_{B,T} \frac{\partial T_{B,T}}{\partial x} + \frac{2h_{B,T}}{R}\left(T_w - T_{B,T}\right) \tag{3.21}$$

where $T_{B,T}(x = 0, t) = T_{B,in}$.

Tube wall:

$$\rho_w C_w \frac{\partial T_w}{\partial t} = \frac{2h_{B,T}}{R}\left(T_w - T_{B,T}\right) - \frac{2h_{T,v}}{R}\left(T_w - T_s\right) \tag{3.22}$$

Net heat flow across tube:

$$Q = \int 2\pi R h_{T,v}\left(T_w - T_s\right)\mathrm{d}x \tag{3.23}$$

Constant parameters:

Holdups	M_{WB1}, M_{WB2}
Densities	ρ_B, ρ_w
Heat capacities	$C_{B,WB1}, C_{B,WB2}, C_{B,T}, C_c, C_w$
Tube dimensions	R, L
Standard temperature	T^*

Inputs:

Brine	$B, T_{B,in}, u_B$
Steam	P_v, H_s
Condensate	W_c
Heat transfer coefficients	$h_{B,T}, h_{T,v}$

With the aforementioned constant parameters and inputs, the model (Equations 3.10, 3.14 through 3.18, and 3.21 through 3.23) would provide a unique solution. PDEs (Equations 3.21 and 3.22) need to be discretized before implementation; a backward finite difference scheme with a parameterized spatial resolution is chosen for this purpose. The discretized model equations constitute an index of one set of DAEs. Feasible initial conditions are any $2NZ + 4$ independent variables where NZ is the number of knots of spatial discretization.

The analogous trajectory for this case for a 5% step increase in the input brine temperature is shown in Figure 3.3. Discretizing the tubes with 60 nodes, the model can produce the hydromechanical dead time fairly well.

Figure 3.4 shows the responses to a step change in steam pressure. As expected, there is no time delay in this case. The time constant is slightly smaller than that found for a step change in the brine temperature. Because the steam phase is modeled quasi-stationary, there is an instantaneous increase in the vapor phase temperature.

3.2.3 Auxiliary Equipment

3.2.3.1 Desuperheater Model

As shown in Figure 3.2, the low-pressure steam is desuperheated by injecting condensate.

The desuperheater is modeled as quasi-stationary by making the following assumptions:

1. The process is adiabatic.
2. Nonthermal energies can be neglected.
3. Holdup is negligible.
4. The residual superheat is perfectly controlled.

The last assumption determines the required condensate flowrate in the simulation of the brine heater performance. However, in the controllability analysis of the entire plant, the condensate flowrate is taken as a manipulated variable.

With the aforementioned assumptions, the desuperheater model equations are as follows.

Mass balance:

$$M_{v,out} = M_{v,in} + M_c \qquad (3.24)$$

Energy balance:

$$M_{v,out}H_{v,out} = M_{v,in}H_{v,in} + M_cH_c \qquad (3.25)$$

Saturation temperature:

$$T_v = T_s(P) \qquad (3.26)$$

Constraint on output steam temperature:

$$\Delta T = \bar{T}_v - T_v \qquad (3.27)$$

3.3 Stage Model

Similar to the steady-state model, the flash stage consists of four compartments (see the representation in Figure 3.5 of the single stage of an MSF plant, namely, the brine pool, product tray, vapor space, and the tube bundle). Now, a generic stage model for dynamic simulation is based on the following considerations.

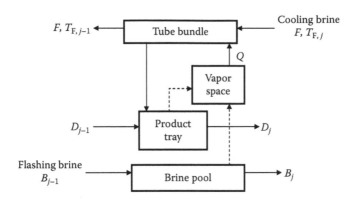

FIGURE 3.5
Representation of a single-stage *j* of the MSF plant.

Uniform mixing of the flashing brine is assumed although certain parameters may remain distributed. Due to a lack of mass transfer data for estimating vaporization rates, nonequilibrium allowances plus other losses are used to relate vapor and brine temperatures. Such allowances are determined using empirical correlations (Lior, 1986). The same are supposed to take care of all the effects due to imperfect mixing, bubble growth, vapor–liquid interface mass transfer, and so on.

The distillate temperature leaving any stage is the same as the vapor condensation temperature. Its flow pattern is that of a well-mixed stirred tank. These assumptions are justified in view of the fact that the vaporization from the distillate normally occurs at its entry into the stage; therefore, its temperature change through the stage is marginal.

The vapor space is considered as lumped stationary, since the percentage of noncondensables in the vapor space is small enough.

Any holdup change for the cooling brine flowing inside the tubes is negligible. Since no evaporation takes place there, the salt concentration remains constant. Hence, the only distributed parameter is the brine temperature. Using the log mean temperature difference as the driving force for heat transfer will not lead to any major error. Moreover, the density and specific heat capacity of the brine inside the tubes are taken as a constant at their average values. Heat losses to the environment are negligible. Changes in potential and kinetic energies are negligible in comparison to changes in thermal energy.

From the aforementioned discussion, it is clear that a flash stage can be adequately represented by four well-mixed stirred tanks for which mass and energy balance equations are written, constituting a set of DAEs.

3.3.1 Flashing Brine

Mass balance:

$$\frac{dM_B}{dt} = B_{in} - B_{out} - V_B \tag{3.28}$$

Salt balance:

$$\frac{d(M_B x_B)}{dt} = B_{in} x_{B,in} - B_{out} x_B \tag{3.29}$$

Enthalpy balance:

$$\frac{d(M_B H_B)}{dt} = B_{in} H_{B,in} - B_{out} H_B - V_B H_{v,B} \tag{3.30}$$

Substituting Equation 3.28 into Equations 3.29 and 3.30 and rearranging:

$$M_B \frac{dx_B}{dt} = B_{in} x_{B,in} - x_B \left(B_{in} - V_B \right) \tag{3.31}$$

$$M_B \frac{dH_B}{dt} = B_{in} H_{B,in} - H_B \left(B_{in} - V_B \right) - V_B H_{V,B} \tag{3.32}$$

The actual and equilibrium flashing brine temperatures are related by non-equilibrium allowances:

$$T_{B-sat} = T_B - T_{NEA} \tag{3.33}$$

The partial pressure of water vapor over the brine is given by

$$P_{B-sat} = P_B (1 - Y_{IM}) \tag{3.34}$$

3.3.2 Distillate Product

Mass balance:

$$\frac{dM_D}{dt} = D_{in} - D_{out} + F_D - V_D \tag{3.35}$$

The distillate leaves the stage in equilibrium with the vapor, therefore

$$T_D = T_{v-sat} = T_{B-sat} - BPE \tag{3.36}$$

Enthalpy balance:

$$D_{in}H_{D,in} + F_D H_D = D_{out}H_D + V_D H_{v,D} \tag{3.37}$$

3.3.3 Vapor Space

Mass balance:

$$\frac{dM_v}{dt} = V_B + V_D + V_{in} + F_I - F_D - V_{out} \tag{3.38}$$

Noncondensable balance:

$$\frac{d(M_v Y_I)}{dt} = F_I + V_{in}Y_{I,in} - V_{out}Y_I \tag{3.39}$$

Substituting Equation 3.38 into Equation 3.39 and rearranging

$$M_v \frac{dY_I}{dt} = F_I + V_{in}Y_{I,in} - Y_I\left(V_B + V_D + F_I + V_{in} - F_D\right) \tag{3.40}$$

due to the demister pressure drop.

$$P_v = P_B - K_{DEM}\frac{u_D^2}{\rho_v} \tag{3.41}$$

Enthalpy balance:

$$V_B H_{v,B} + V_D H_{v,D} + V_{in}H_{v,in} + F_I H_I = F_D H_D + V_{out}H_v \tag{3.42}$$

Partial pressure of water vapor in the vapor space:

$$P_{v-sat} = P_v(1 - Y_{IM}) \tag{3.43}$$

3.3.4 Cooling Brine

Mass balance:

$$F_{in} = F_{out} = F \tag{3.44}$$

Salt balance:

$$F_{in}x_{F,in} = F_{out}x_{F,out} \tag{3.45}$$

Enthalpy balance:

$$M_F C_F \frac{dT_{F,out}}{dt} = F C_F \left(T_{F,in} - T_{F,out} \right) + Q \tag{3.46}$$

Heat transfer:

$$Q = U A \Delta T \tag{3.47}$$

where

$$\Delta T = \frac{T_{F,out} - T_{F,in}}{\ln \left(\dfrac{T_D - T_{F,in}}{T_D - T_{F,out}} \right)} \tag{3.48}$$

Condensation rate:

$$F_D = \frac{Q}{H_v - H_D} \tag{3.49}$$

Note the following:

1. F_I, the flowrate of noncondensables from the brine pool, being small, is not shown separately in the mass balance equation (Equation 3.28).
2. Temperature loss due to pressure drop in the demister is also very small; hence, it is not included in Equation 3.36.

Thus, Equations 3.31 through 3.37 and 3.40 through 3.49 constitute a set of DAEs, which represent a dynamic model for any stage j in the heat recovery or rejection section. In the first stage ($j = 1$), no distillate flow enters; therefore, D_{in} and V_D should be equated to zero. The last stage ($j = N$) receives make-up flow F_m; the recycle R and blowdown BD are taken out. Consequently, Equations 3.28, 3.31, and 3.32 are modified as follows:

$$\frac{dM_B}{dt} = B_{in} + F_m - R - BD - V_B \tag{3.28a}$$

$$M_B \frac{dx_B}{dt} = B_{in} x_{B,in} + F_m x_{sea} - x_B \left(B_{in} + F_m - V_B \right) \tag{3.31a}$$

$$M_B \frac{dH_B}{dt} = B_{in} H_{B,in} + F_m H_m - H_B \left(B_{in} + F_m - V_B \right) - V_B H_{v,B} \tag{3.32a}$$

3.4 Interstage Flow

The hydrodynamics of the flash stage in the MSF plant is influenced by the interstage flow arrangements and stage internals. Sluice gate orifices are most commonly used for flow regulation between the stages. The flashing brine entering a stage through the orifice from an upstream stage is divided, due to the partition wall supports, into as many streams as the number of supports. These emerging jets are either submerged or nonsubmerged based on the tail water depth, which is the liquid depth at the exit of the stage. If the jet discharges into the free space, it is called *free discharge* or *nonsubmerged flow*. On the other hand, if it discharges into a liquid pool, it is known as submerged flow. In both the cases, the jet contracts until it reaches the vena contracta, in which the paths of all the elements of the jet are parallel and the pressure in the jet can be assumed to be equal to that in the surrounding fluid.

For most of the orifice shapes, the vena contracta forms at some distance downstream of the orifice with contractions all around. If the edge of the orifice is flushed with the wall or floor, contraction on that side will be entirely eliminated. However, there is no change in the contraction on the other side. Rounding the inner edge of the orifice reduces the contraction, which ultimately can be eliminated by shaping the orifice to conform with the form of the contracting jet. The contraction coefficient is defined as the ratio of the flow area at the vena contracta to the orifice area. The following equation is given by Hoemig (1978) to evaluate the contraction coefficient:

$$C_c = 0.61 + 0.18\bar{X} - 0.58\bar{X}^2 + 0.7\bar{X}^3 \tag{3.50}$$

where

$$\bar{X} = \frac{h_o}{\left[\dfrac{(P_1 - P_2)}{\rho} + L_1 \right]} \tag{3.51}$$

3.4.1 Stages with Kick-Plates

In this case, the vena contracta forms in both the horizontal and vertical directions at a distance equal to the orifice opening downstream. Then, these jets expand further down and form a single sheet roughly at a distance of five orifice openings. The stream lines are parallel at the vena contracta, and the flow can be supercritical or subcritical. This depends upon the value of the Froude number $\left(Fr = u/\sqrt{gL} \right)$ for the flow conditions at the vena contracta: for $Fr = 0$ critical flow, $Fr < 1$ subcritical, and $Fr > 1$ supercritical.

In full-scale MSF plants, in most stages, the flow at the vena contracta is supercritical except for those in the heat rejection section. In the absence of a kick-plate, the brine jet will pass through the stage creating blowthrough conditions, which is not desirable. If the kick-plate is there, the supercritical flow would either change into subcritical accompanied by a hydraulic jump or pass as such, depending upon the kick-plate height and the liquid level over the kick-plate. The height, shape, and toe of the hydraulic jump are functions of Fr of the approaching stream, height of the kick-plate, and liquid depth over it. The toe of the jump can form before, on, or past the kick-plate. If the flow is subcritical, then depending on the kick-plate height, it may turn into critical flow downstream of the kick-plate; in such a case, it would affect the upstream levels or form a liquid hump over the kick-plate.

Furthermore, depending on the orifice as well as kick-plate height, two hydraulic jumps may form: one downstream of the orifice and the other downstream of the kick-plate. These hydraulic jumps no doubt help in better mixing in the stage and sealing the orifice to avoid blowthrough conditions.

The basic criterion for the hydraulic jump to occur is that the jet from the orifice should be supercritical and the kick-plate height sufficient to promote the jump. The toe of the jump in all the stages is before the kick-plate, and in some stages, it forms upstream of the vena contracta. An increase in the brine level just before the kick-plate changes the flow regime from nonsubmerged to submerged flow.

For the estimation of the liquid level above the kick-plate, the following weir flow equations can be used (Ishihara and Ida, 1951; Villemonte, 1947):

$$B = K_2 w_s \bar{w} \rho \tag{3.52}$$

where

$$\bar{w} = K_1 (L_2 - h_w)^{1.5} \tag{3.53}$$

$$K_1 = \frac{1.785 + 0.00295}{(L_2 - h_w)} + \frac{0.237(L_2 - h_w)}{h_w} \tag{3.54}$$

For the submerged weir, $L_3 > h_w$

$$K_2 = \left(1 - \left[\frac{L_3 - h_w}{L_2 - h_w}\right]^{1.5}\right)^{0.385} \tag{3.55}$$

For the crested weir, $L_3 > h_w$ and $K_2 = 1.0$.

3.4.1.1 Level at the Vena Contracta

The brine level at the vena contracta affects the brine flow through the orifice. For nonsubmerged flow, this level is the jet height, while in the submerged flow, the brine level is more than the jet height of the brine. There are considerable energy losses in the hydraulic jump, which cannot be estimated accurately. A substantial amount of energy is lost due to friction between the vena contracta and kick-plate. Therefore, an energy balance between these two points cannot provide a correct estimation of the liquid level at the vena contracta. A better alternative is the momentum balance as it does not contain any energy loss term. In making this balance, forces offered by the floor friction are neglected as they are small. The momentum balance for flow in a rectangular open channel is written as follows:

$$(w_sL)\left(\frac{\rho g L}{2}\right) + Bu_v = (w_sL_2)\left(\frac{\rho g L_2}{2}\right) + Bu_2 \tag{3.56}$$

where

$$u_v = \frac{B}{(\rho w_o C_c h_o)} \tag{3.57}$$

$$u_2 = \frac{B}{(\rho w_s L_2)} \tag{3.58}$$

On the left-hand side of Equation 3.56, the first term represents momentum due to hydrostatic pressure at the vena contracta and the second term momentum is due to jet velocity. On the right-hand side, the first term gives the momentum due to the hydrostatic head just upstream of the kick-plate and the second term momentum is due to the velocity of liquid just upstream of the kick-plate. Now, substituting Equations 3.57 and 3.58 into Equation 3.56 and solving for L,

$$L = \left(\frac{2A}{gL_2} + L_2^2 - \frac{2\bar{A}}{gC_c h_o}\right)^{0.5} \tag{3.59}$$

where

$$A = \left[\frac{B}{\rho w_s}\right]^2; \quad \bar{A} = \frac{(B/\rho)^2}{(w_s w_o)}$$

The brine level estimated by Equation 3.59 helps to identify the flow regimes: if the brine level is less than or equal to the jet height, the flow is nonsubmerged; otherwise, it is submerged.

In the case where the flow is subcritical at the vena contracta, as expected in the heat rejection section, no hydraulic jump occurs. Instead, backflow occurs since the weir is sufficiently high. The same momentum balance equation as described earlier for the supercritical flow with hydraulic jump will provide the liquid level at the vena contracta. For both the supercritical and subcritical approach flows, there is no possibility of bump formation over the weir, since the weir height is more than the minimum in the commercial MSF plants.

3.4.1.2 Level Upstream of the Orifice

From a stage to its downstream stage, the brine flows due to the differences in the brine level and pressure on both sides of the orifice. At the upstream side of the orifice, the flow is subcritical and normally changes to supercritical at the downstream. This transaction is smooth and the head loss is about 4%–5% of the total head depending upon the flowrate, orifice type, and opening. At the vena contracta, stream lines are parallel; if a point is selected slightly away from the orifice at the upstream, the energy balance can be written as

$$L_1 + \frac{u_1^2}{2g} + \frac{P_1}{\rho g} = L + \frac{u_v^2}{2g} + \frac{P_2}{\rho g} \tag{3.60}$$

where

$$u_1 = \frac{B}{(\rho w_s L_1)} \tag{3.61}$$

Substituting Equations 3.57 and 3.61 into Equation 3.60 and rearranging

$$2gL_1^3 + DL_1^2 + A = 0 \tag{3.62}$$

where

$$D = \frac{2(P_1 - P_2)}{\rho} - 2gL - \frac{\hat{A}}{C_c^2 h_o^2}$$

$$A = \left[\frac{B}{\rho w_s} \right]^2 ; \quad \hat{A} = \left[\frac{B}{\rho w_o} \right]^2$$

Equation 3.62 is a cubic equation in L_1, which can be solved using the procedure given in Perry and Green (1984). For the operating conditions in the MSF plant, two positive roots and one negative root are obtained for L_1. For the submerged flow, a large positive root should be taken for L_1, since the small positive root for L_1 is the sequent depth to L_1. For the nonsubmerged flow, the small positive root should be assigned to L_1 as the larger one is the sequent depth to L_1. For estimating brine flow for a specific level, Equation 3.60 can be rearranged to

$$B = \left[\frac{2g\rho^2 \left(L_1 - L + \dfrac{P_1}{\rho g} - \dfrac{P_2}{\rho g} \right)}{\dfrac{1}{\left(w_o^2 C_c^2 h_o^2 \right)} - \dfrac{1}{\left(w_s^2 L_1^2 \right)}} \right]^{0.5} \tag{3.63}$$

3.4.2 Brine Holdup

For the submerged flow, the brine level at the vena contracta and the level at the end of the stage do not differ much. Hence, the stage holdup estimated by using the average of these two levels would be sufficiently accurate.

In the case of nonsubmerged flow, brine jets between the stage partition supports combine at a short distance from the orifice, and then flow for a certain distance before the hydraulic jump occurs. Subsequently, brine flows up to the end of the stage almost at the level equal to the jump level. For the nonsubmerged flow, therefore, the stage is divided into three distinct regions, namely: (1) prehydraulic jump, (2) jump region, and (3) posthydraulic jump.

For the estimation of holdup in this case, the level profiles and demarcation between the above regions are necessary.

Under the operating conditions of a large-scale MSF desalination plant, the hydraulic jump generally forms before the kick-plate and ends over the kick-plate. For the nonsubmerged flow, the jump length determines the location of the toe of the jump. However, the jump length cannot be easily predicted by theory, but it has been investigated experimentally. Chow (1959) fitted the following correlations in different ranges of Froude numbers by plotting the experimental data in the form of a ratio of jump length to jump height versus the approach flow Fr:

$$\left. \begin{aligned} &X_1/L_2 = 2.2333 + 1.5 Fr; \quad 0.5 \le Fr \le 1.5 \\[4pt] &X_1/L_2 = 1.2544 + 2.0859 Fr - 0.2854 Fr^2 + 0.0119 Fr^3; \quad 1.5 \le Fr \le 4.5 \\[4pt] &X_1/L_2 = 3.936 + 0.6871 Fr - 0.0543 Fr^2; \quad 4.5 \le Fr \le 6.5 \\[4pt] &X_1/L_2 = 5.6295 + 0.1619 Fr - 0.019027 Fr^2 + 0.000981 Fr^3 \\[4pt] &\qquad -0.0000227 Fr^4; \quad 5.5 \le Fr \le 20 \end{aligned} \right\} \tag{3.64}$$

The aforementioned correlations for estimating the hydraulic jump are determined in the absence of flashing. Abdelmassih and Hsu (1976) studied the effect of flashing rate on the shape and size of the hydraulic jump using tap water in their experiments; only qualitative information is given. Salts dissolved in seawater can also affect the characteristics of hydraulic jump due to foaming, but such information is not available.

3.4.2.1 Hydraulic Jump Profile

For pure water flow without flashing, the following relation has been suggested by Subramanya (1982) to evaluate brine holdup in a case of nonsubmerged jump:

$$\bar{L} = 0.75(L_2 - L_v)\eta \tag{3.65}$$

The parameter η in Equation 3.65 is a function of $\lambda = x/X_h$, where x is the required horizontal position from toe to the jump and X_h is given by

$$X_h = L_v(5.08Fr - 7.82) \tag{3.66}$$

η is correlated in terms of λ, as follows:

$$\left.\begin{aligned}
\eta &= 5.6167\lambda - 45.0\lambda^2 + 1333.333\lambda^3; \quad 0 \le \lambda \le 0.15 \\[4pt]
\eta &= 0.1543737 + 0.8317837\lambda; \quad 0.15 \le \lambda \le 0.8 \\[4pt]
\eta &= -2.9239 + 11.0011\lambda - 12.5922\lambda^2 + 7.461723\lambda^3 - 2.20375\lambda^4 \\[4pt]
&\quad + 0.255568\lambda^5; \quad 0.8 \le \lambda \le 2.4
\end{aligned}\right\} \tag{3.67}$$

Using Equation 3.65, Simpson's numerical integration procedure can be applied to evaluate the brine holdup.

3.4.2.2 Region of Gradually Varied Flow

The region between the orifice and toe of the hydraulic jump has a gradually varied flow in open channel hydraulics. Most of the flash evaporation occurs in this region with a two-phase flow prevailing. The brine level profiles in the said region can be derived as follows, starting with the energy equation for a horizontal channel. The specific energy is written as

$$H = y + \frac{u^2}{2g} \tag{3.68}$$

where
  H is the energy head (m)
  y is the brine level (m)
  u is the flow velocity (m/s)

Differentiating and rearranging Equation 3.68,

$$\frac{dy}{dx} = \frac{-s}{1 + \dfrac{d}{dy}\left(\dfrac{u^2}{2g}\right)}$$

(3.69)

where energy slope is $-dH/dx = s$.
 The energy slope (s) is estimated from Manning's formula

$$s = \frac{n^2 u^2}{R^{4/3}}$$

(3.70)

where n is Manning's constant.
 The hydraulic radius R for the rectangular channel is written as

$$R = \frac{y w_s}{(w_s + 2y)}$$

(3.71)

and the flow velocity in a rectangular channel

$$u = \frac{B}{(w_s \rho y)}$$

(3.72)

Substituting Equations 3.70 through 3.72 into Equation 3.69 and rearranging

$$\left(\frac{B^2 y^{1/3}}{\rho^2 w_s^2 g (w_s + 2y)^{4/3}} - \frac{y^{10/3}}{(w_s + 2y)^{4/3}}\right) dy = \frac{B^2 n^2}{\rho^2 w_s^{10/3}} dx$$

(3.73)

in the nonsubmerged flow $y \ll w_s$; therefore, Equation 3.73 can be simplified to

$$\left(\frac{B^2 y^{1/3}}{\rho^2 g w_s^{10/3}} - \frac{y^{10/3}}{w_s^{4/3}}\right) dy = \frac{B^2 n^2}{\rho^2 w_s^{10/3}} dx$$

(3.74)

Rearranging and integrating Equation 3.74,

$$x = \left(\frac{3}{4}\right)\frac{y^{3/4}}{n^2 g} - \left(\frac{3}{13}\right)\frac{1}{n^2}\left(\frac{\rho w_s}{B}\right)^2 y^{13/3} + \text{constant}$$

(3.75)

For smooth metals, the value of n is 0.01–0.012 in SI units. Putting $n = 0.012$ in Equation 3.75

$$x = 530.92 y^{4/3} - 1602.56\left(\frac{\rho w_s}{B}\right)^2 y^{13/3} + \text{constant}$$

(3.76)

In certain MSF plants, all flash stages are divided and installed in two decks, one over the other. The last stages of both the top and bottom decks are not provided with any kick-plates. In the last stage of the bottom deck, the brine level is controlled at a certain value to avoid cavitation of the blowdown and recycle pumps; therefore, there is always a submerged flow. From the last stage of the upper deck, the brine falls into a small channel, which acts as a liquid seal to avoid blowthrough. Hence, no hydraulic jump occurs and the flow is gradually varied in this stage and the brine level can be predicted by Equation 3.76.

3.4.3 Holdup Calculation in Dynamic Simulation

In the dynamic simulation of the MSF desalination plant, brine holdup in each stage is to be calculated for any disturbance at every time step in the numerical integration. Using the equations as given earlier for the stages supplied with kick-plates, the following stepwise procedure has been used by Reddy et al. (1995b):

Step 1: Start with the brine level at the vena contracta and at the exit of each stage available from the steady-state simulation or from the previous integration step.

Step 2: Calculate the brine flowrate from Equation 3.63.

Step 3: Estimate the brine level at the kick-plate using the weir equations (Equations 3.52 through 3.55).

Step 4: Calculate the brine level at the vena contracta from Equation 3.59. If the present and previous levels at the vena contracta agree within a tolerable limit, go to the next step; otherwise, repeat from step 2 with an updated level.

Step 5: For stages with submerged flow, estimate the stage holdup from the average of the levels at the exit and vena contracta. For stages with nonsubmerged flow, calculate the length of the hydraulic jump using the appropriate correlation among Equation 3.64. Calculate the toe of the jump assuming that the jump ends on the kick-plate. Estimate brine holdup using Equation 3.65 for the hydraulic jump and Equation 3.75 for the region from the orifice to the toe of the hydraulic jump. For the region from the kick-plate to the exit of the orifice, estimate the holdup with the average of the levels at both the ends.

Step 6: Check the estimated holdups with the holdups obtained from integration. If they match within tolerable limits for each stage, proceed to the next integration step. Otherwise, update the stage exit levels with new holdups and repeat from step 1.

In step 1, steady-state brine levels may be required to continue the stepwise calculation. Hence, the following iterative procedure to estimate brine levels in the steady-state simulation is given by Reddy et al. (1995b).

3.4.3.1 Steady-State Simulation

As already mentioned, the brine level in the last MSF stage is controlled to avoid cavitation in the process. Thus, in this stage, the flow is uniform and submerged. Using this level, the following procedure has to be followed to compute the level in the next upstream stage:

1. Calculate the level at the exit of the stage by using Equation 3.62.
2. Calculate the brine level just upstream of the kick-plate from Equation 3.52 using the tail brine level obtained from step 1.
3. Estimate the vena contracta brine height with the contraction coefficient calculated from Equation 3.50, neglecting the horizontal contraction. Calculate Fr at the vena contracta.
4. Calculate the brine level at the vena contracta using the momentum balance equation (Equation 3.59). If it is less than the vena contracta height obtained in step 3, equate it to the vena contract height and the resulting flow is nonsubmerged. Otherwise, the flow is submerged and the prevailing brine level is the brine level at the vena contracta calculated in the current step.
5. For the submerged flow, estimate the stage holdup using the average of the brine levels at the exit of the stage and vena contracta.
6. For the nonsubmerged flow, calculate the length of the hydraulic jump using the appropriate equation (Equation 3.64). Calculate the toe of the jump assuming that the jump ends on the kick-plate. Estimate the brine holdup using Equations 3.65 and 3.75 for the regions of the hydraulic jump and orifice to toe of the hydraulic jump, respectively. For the region between the kick-plate and exit of the orifice, estimate the holdup with the average of the levels at both the ends.
7. Perform the computation from steps 1 to 6 for all the bottom deck stages using the brine level at the vena contracta of the downstream stage.
8. For the last stage in the upper deck, estimate the brine-level profile and holdup using Equation 3.75.
9. Calculate the brine levels and holdups in all the stages upstream of the last stage in the upper deck following steps 1–6.
10. Compute material and energy balances for each stage with updated brine levels. Repeat the calculations till the brine levels converge.

3.4.4 Noncondensable Gases

Atmospheric gases such as oxygen, nitrogen, and argon are molecularly dissolved in the seawater and liberated mainly in the deaerator. On the other hand, carbon dioxide reacts chemically with the seawater to form carbonic acid that, in turn, dissociates into bicarbonate and carbonate ions. In acid-treated plants, CO_2 is removed in the carbonator. On the other hand, in the additive-treated MSF plants, CO_2 is released in the flash chambers and should be extracted adequately by venting; otherwise, heat transfer rates will be significantly reduced. Due to their large Henry coefficients, noncondensable gases (NC) make an appreciable contribution to the vapor space pressure. Moreover, CO_2 and oxygen corrode the shell side of the condensers and lead to tube leakages.

A great deal of uncertainty prevails in determining the release rates of CO_2 in the MSF distillers due to a lack of knowledge about the kinetics of the chemical reactions involved, as well as the influence of the mass transfer process. For that reason, design information varies widely as far as CO_2 release rates are concerned.

Seifert (1988) proposed a semiempirical model for the release of NC gases in MSF plants, in which the main emphasis was on the mass transfer resistance hindering the release of such gases. The model proposed by Genthner and Seifert (1991) included the dissociation of water and CO_2 according to the following equilibrium reactions:

$$H_2O \rightleftarrows OH^- + H^+ \tag{3.77}$$

$$CO_2 + H_2O \rightleftarrows H^+ + HCO_3^- \tag{3.78}$$

$$HCO_3^- \rightleftarrows H^+ + CO_3^{2-} \tag{3.79}$$

In their model, all other species were represented in terms of the ionic strength of the solution and not accounted for individually. The activities of all ionic species were approximated by the Debye–Huckel equation and equilibrium constants by temperature-dependent correlations. The ionic strength was determined by using an empirical expression in terms of the total dissolved solids. Scale formation and solid precipitation were not considered. Upon implementation of the Seifert–Genthner model, Marquardt (1996c) noted several inconsistencies; therefore, he used a different set of correlations (Hancke, 1994) to calculate the equilibrium constants.

According to present knowledge, reactions (3.78) and (3.79) describe the CO_2/ seawater system at lower temperatures only. At higher temperatures, because of the evolution of molecular CO, the equilibrium between CO_2, HCO_3^-, and CO_3^{2-} is disrupted. Moreover, additional CO_2 is formed due to thermally induced reactions. The primary reaction, which leads to further evolution of CO_2 and triggers alkaline scale formation (Glater et al., 1980), is the thermal decomposition of HCO_3^- ions, for which Langelier et al. (1950) suggested the following mechanism:

$$2HCO_3^- \rightleftarrows CO_2 + CO_3^{2-} + H_2O \tag{3.80}$$

The carbonate ions, thus generated, can cause precipitation of calcium carbonate once its solubility limit is exceeded. At still higher temperatures, CO_3^{2-} ions may be partially or totally hydrolyzed and, as a result, the concentration of OH^- ions increases, leading to the precipitation of magnesium hydroxide if sufficient Mg^{2+} ions are available in the solution to satisfy the solubility limit. In the case of total hydrolysis of CO_3^{2-} ions, more molecular CO_2 will form as follows:

$$CO_3^{2-} + H_2O \rightleftarrows CO_2 + 2OH^- \qquad (3.81)$$

Dooly and Glater (1972) suggested unimolecular decomposition of HCO_3^- instead of a bimolecular reaction (3.4) according to

$$HCO_3^- \rightleftarrows CO_2 + OH^- \qquad (3.82)$$

However, the investigations carried out by Shams et al. (1989a) confirmed the bimolecular decomposition of HCO_3^- according to reaction (3.80) and complete hydrolysis of CO_3^{2-} according to reaction (3.81).

The release of CO_2 in the MSF evaporators is thus influenced by the following factors:

1. Prevailing temperature and pressure profiles, particularly the maximum temperature
2. Reaction kinetics, mass transfer rate, and, hence, residence time
3. The HCO_3^- and CO_3^{2-} content of the makeup stream
4. Deaeration effects, such as agitation of the brine
5. The availability of Ca^{2+} and Mg^{2+} ions
6. The presence of antiscalants

In the various approaches proposed to compute CO_2 release rates in MSF distillers, the relative importance attached to these parameters is controversial. The models suggested differ not only in the procedure for computing the total release rate of CO_2 but also its distribution among the individual stages.

The Ciba–Geigy (1978) model is based on reaction (3.80) for the CO_2 release rate, assuming that the hydrogen carbonate concentration in the seawater is approximately equal to the value of the total alkalinity (ppm $CaCO_3$ of the makeup flow). Thus, this model does not take into account (1) the hydrolysis of CO_3^{2-} at higher temperatures in addition to HCO_3^- decomposition and (2) any effect of temperature, reaction kinetics, and mass transfer on the extent of decomposition reaction and, hence, the residual concentration of HCO_3^- ions in the brine of the last stage.

A semiempirical model given by Watson Desalination Consultants (1979) for computing the CO_2 release rate is based on HCO_3^- measurements in the last stage, in which dependence of the HCO_3^- decomposition on the brine residence time is implicitly involved. In this model, CO_2 diffusion is not the determining factor in its release, which may be hindered by the reaction kinetics or mass transfer process. The Watson Report further assumed the following stage-wise pattern in the high-temperature stages for the release of CO_2 from the thermal decomposition of HCO_3^- : stage 1, 85%; stage 2, 10%; stage 3, 5%; and virtually, no CO_2 is released from the fourth stage onward.

Seifert's (1988) investigation showed that the chemical reactions occurring in the CO_2/seawater system are faster than the mass transfer rate of CO_2 from the brine to the vapor. Therefore, the transfer resistance of CO_2 in the interface between brine and vapor is the determining factor in the CO_2 release rate.

In a study based on electrolyte equilibria, Marquardt (1996b) found that the CO_2 formation rate depends strongly on the operating conditions. A reduction in the TBT leads to a significant decrease in the CO_2 formation rate in the first few stages and a corresponding increase in the subsequent stages. The influence of the vapor flowrate is more involved; a reduction in the vapor flowrate results in a corresponding reduction in the CO_2 formation rate in the first stage only, while for all other stages, it is enhanced. As reported by Marquardt (1996), almost all of the CO_2 is formed and released in the first four flash stages.

3.4.5 Modeling Venting System

A typical venting system is shown in Figure 3.6, which consists of two generic components, namely, ejectors and vent condensers. For both the equipment items, generic models are derived.

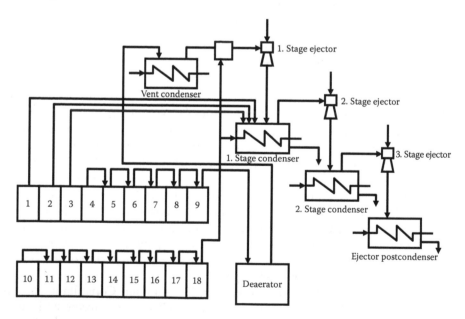

FIGURE 3.6
Venting system of a typical MSF desalination plant.

3.4.5.1 Ejector Model

An ejector is characterized by complex thermodynamic and fluid mechanic interactions. In the ejector, motive steam is expanded into a supersonic jet, entraining a certain quantity of gas through its suction chamber. In modeling the ejectors, the following assumptions are justified:

1. Adiabatic operation
2. No entrained gas in motive steam
3. Negligible residence time
4. Characteristic curve represents ejector performance

Since the operation is adiabatic, the ejector functions as a mixer in terms of heat and mass balances. With a small surface area and low heat transfer coefficients, heat losses to the environment can be neglected. All capacitances are ignored, since residence time is extremely short due to high velocity and low density of the fluid.

The model equations are as follows:

Mass balance:

$$V_{\text{in}} + V_{\text{ms}} = V_{\text{out}} \tag{3.83}$$

Component balance:

$$\left.\begin{aligned} V_{\text{in}} y_{\text{NC,in}} &= V_{\text{out}} y_{\text{NC,out}} \\ V_{\text{in}} y_{\text{w,in}} + V_{\text{ms}} &= V_{\text{out}} y_{\text{w,out}} \end{aligned}\right\} \tag{3.84}$$

Energy balance:

$$V_{\text{in}} H_{\text{v,in}} + V_{\text{ms}} H_{\text{ms}} = V_{\text{out}} H_{\text{v,out}} \tag{3.85}$$

where
  V denotes the molar flowrate
  y denotes the mole fraction
  the subscript ms denotes the motive steam

The molar specific enthalpies are provided by suitable property routines:

$$\left.\begin{aligned} H &= H(T,P,y) \\ H_{\text{ms}} &= H_{\text{ms}}(T,P) \end{aligned}\right\} \tag{3.86}$$

In addition to these balance equations, phenomenological relations are needed to relate the discharge pressure to the suction pressure, flowrate, and pressure of the motive steam. Because of the complex phenomena involving supersonic flow, compression shocks, and high turbulence, only a parametric model can be applied. A characteristic curve from Perry and Green (1984) is linearized in the vicinity of the normal operating point, which provides the following expressions for the flowrate of the entrained gas V_{in} and the discharge pressure P_{out}:

$$\left.\begin{array}{l} V_{in} = V_{ms} \dfrac{MW_{ms}}{MW_{in}} \left[\dfrac{V_{in}}{V_{ms}}\right]^* \left[\dfrac{(P_{in}/P_{ms})}{(P_{in}/P_{ms})^*}\right]^a \\[20pt] P_{out} = P_{in} \left[\dfrac{P_{out}}{P_{in}}\right]^* \left[\dfrac{(P_{out}/P_{ms})}{(P_{out}/P_{ms})^*}\right]^b \end{array}\right\} \qquad (3.87)$$

In these equations, the expressions marked with an asterisk pertain to the nominal operating point, and MW is the molecular weight. MW_{in} is calculated as follows:

$$MW_{in} = y_{w,in} MW_{ms} + y_{NC,in} MW_{NC} \qquad (3.88)$$

Using the design data, the various parameters as calculated are given in Table 3.1. The following quantities are time-invariant parameters of the system. The following are taken as known inputs to the system:

V_{ms}	Motive steam flowrate (kmol/s)
P_{ms}	Motive steam pressure (bar)
$H_{V,in}$	Molar enthalpy of entrained gas (mJ k/mol)
$y_{NC,in}$	Composition of entrained gas (mole fraction)
P_{in}	Suction pressure (bar)

With these quantities known or fixed, the given set of algebraic equations can be solved to find the remaining unknowns.

TABLE 3.1

Design Data and Derived Parameters for Ejector Models

	Ejector 1	Ejector 2	Ejector 3
Suction pressure (mbar)	71.2	135	298
Discharge pressure (mbar)	145	308	1018
Motive steam pressure (bar)	16	16	16
Motive steam flowrate (kg/s)	1093	393	404
$(v_{in}/v_{ms})^*$	0.72	1.1	0.523
$(P_{out}/P_{ms})^*$	0.001	0.0192	0.064
$(P_{out}/P_{in})^*$	2.03	2.28	3.41
Parameter a (Equation 3.87)	1.0744	1.0744	1.6162
Parameter b (Equation 3.87)	−0.6424	−0.6264	−0.5508

Note: $(P_{out}/P_{ms})^*$, nominal ratios, discharge to motive steam pressure; $(P_{out}/P_{in})^*$, nominal compression ratios; $(V_{in}/V_{ms})^*$, nominal ratio between flowrates of entrained gas and motive steam.

3.4.5.2 Vent Condenser Model

The vent condensers are shell and tube heat exchangers with vapor condensing on the outside surface of the tube bundle and makeup water flows as coolant inside the tubes. The vapor–liquid equilibrium relationship will have to be considered here in modeling the vent condenser, since the predominant constituent of the noncondensable gases, that is, CO_2, will be present in both the liquid and vapor phases. For the sake of simplicity, the total noncondensable component going to the condenser can be identified as carbon dioxide only. Hence, modeling of the vent condenser is based on the following assumptions:

1. Only water and carbon dioxide are present in the venting system.
2. Both the liquid and vapor phases are well mixed.
3. Condensate and outlet vapor approach equilibrium conditions.
4. Residence time is negligible compared to plant time constants.
5. The arithmetic mean temperature difference can be used in heat exchanger calculations.

Due to the short residence time, the balance equations are written as steady-state balances.

3.4.5.2.1 Shell Side

Mass balance:

$$V_{\text{in}} = V_{\text{out}} + F_{\text{con}} \tag{3.89}$$

Component balance:

$$V_{\text{in}} y_{i,\text{in}} = V_{\text{out}} y_{i,\text{out}} + F_{\text{con}} x_{i,\text{con}}; \quad i \in CO_2, \text{w} \tag{3.90}$$

Energy balance:

$$V_{\text{in}} H_{v,\text{in}} = V_{\text{out}} H_{v,\text{out}} + F_{\text{con}} H_{\text{con}} + Q \tag{3.91}$$

Equilibrium relationship:

$$y_{i,\text{out}} = K_i x_{i,\text{con}}; \quad i \in CO_2, \text{w} \tag{3.92}$$

Vapor flow (orifice flow relation):

$$V_{\text{out}}^2 = k(P - P_{\text{out}}) \tag{3.93}$$

Thermodynamic properties:

$$\left.\begin{aligned}
H_{v,out} &= H_v(T,P,y) \\
H_{con} &= H_L(T,P,x) \\
K_i &= K_i(T,P,x_{i,con})
\end{aligned}\right\} \quad (3.94)$$

The pressure P_{out} in Equation 3.93 is the pressure at the suction chamber of the adjacent ejector downstream of the condenser.

3.4.5.2.2 Tube Side
Mass balance:

$$F_{in} = F_{out} = F \quad (3.95)$$

Energy balance:

$$FC_F\left(T_{F,out} - T_{F,in}\right) = Q \quad (3.96)$$

Heat transfer:

$$Q = UA\left[T_v - \left(\frac{T_{F,in} + T_{F,out}}{2}\right)\right] \quad (3.97)$$

The following are the time-invariant parameters of the model:

U	Average heat transfer coefficient (W/m² K)
A	Condenser heat transfer area (m²)
C_F	Specific heat capacity of the coolant (mJ k/mol K)
k	Orifice coefficient (kmol²/s² bar)

The following are the known inputs:

V_{in}	Input vapor flowrate (kmol/s)
$H_{v,in}$	Input vapor molar enthalpy (mJ k/mol)
F	Coolant flowrate (kmol/s)
$T_{F,in}$	Input liquid temperature (°C)
P_{out}	Suction chamber pressure of adjacent ejector (bar)

Given these quantities, the set of algebraic model equations can be solved to find remaining unknowns. The vent gas composition, $y_{i,out}$, needs special consideration. Depending upon the location of the vent point, a certain composition of the vent gas is obtained. Here, two limiting cases can be

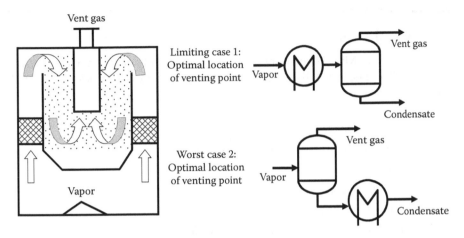

FIGURE 3.7
Vent point geometry and limiting cases for vent gas composition.

distinguished as shown in Figure 3.7. In the limiting case 1, the venting point is so located that the vapor is first cooled to the saturation temperature before noncondensable gas is separated. Due to the vapor–liquid equilibrium, in this case, the maximum concentration of the noncondensable gas is obtained with minimum loss of water vapor. In the limiting case 2, the gas stream is first taken out prior to any cooling and condensation; here the vapor will have a minimum concentration of noncondensable gas and the plant will suffer from a maximum loss of water vapor. The model equations giving the vent gas composition for both cases follow.

$$\text{Case 1:} \quad \begin{cases} y_{w,out}^{opt} = \left(V_{in}y_{w,in} - F_{con}\right)/V_{out} \\ y_{CO_2,out}^{opt} = \left(1 - y_{w,out}^{opt}\right) \end{cases} \tag{3.98}$$

$$\text{Case 2:} \quad \begin{cases} y_{w,out}^{worst} = y_{w,in} \\ y_{CO_2,out}^{worst} = y_{CO_2,in} \end{cases} \tag{3.99}$$

In practice, however, some intermediate value, $y_{i,out}$, will be obtained between the two limiting cases, depending on the heat exchanger design and operating conditions. Therefore, the vent gas composition is better calculated by the weighted average of the two limiting cases by introducing an adjustable venting efficiency η:

$$y_{i,out} = \eta y_{i,out}^{opt} + (1 - \eta)y_{i,out}^{worst} \tag{3.100}$$

3.5 Control Loops

A conventional MSF plant contains several control loops that are shown in Figure 2.4. These are listed in the following, each is provided with a PI controller.

1. TBT control loop
2. Brine heater steam temperature control loop
3. Steam condensate-level control loop
4. Culvert flow control loop
5. Reject seawater control loop
6. Cooling seawater temperature control loop
7. Distillate-level control loop
8. Brine-level control loop (last stage)
9. Brine recycle flow control loop
10. Makeup flow control loop
11. Antiscale flow control loop
12. Sodium sulfide flow control loop
13. Lime flow control loop

Out of these control loops, the most important one is the control of TBT at the brine heater outlet. Fluctuations in TBT are the consequence of disturbances in the heating steam supply (i.e., pressure/flowrate) and/or in the brine recycle flowrate. The efficiency of the brine heater control thus plays an important role in reestablishing a stable operation with the shortest possible delay. Therefore, it is important for the large MSF plant to know which control scheme should be adopted for the brine heater. This problem will be illustrated in the following section by simulating the dynamic behavior of different brine heater control systems.

3.5.1 Valve Model

In addition to the models described earlier under the brine heater model, valve and controller models will be required to simulate the TBT control loop. Butterfly valves are normally used in large-scale MSF plants for flow control due to their inherent equal percentage characteristic. These are rotary motion valves with rotating disks. The valve model describes the flow characteristics of the valve and the inertia of the valve drive. The following equations are written for the steady-state behavior.

Mass balance:

$$W_{s,in} = W_{s,out} \qquad (3.101)$$

Energy balance:

$$W_{s,in}H_{s,in} = W_{s,out}H_{s,out} \qquad (3.102)$$

Flow characteristic:

$$W_s = k_{vs} \cdot C_G \sqrt{\frac{P_{out} \cdot \Delta P \cdot \rho_s}{T_s}} \cdot \phi \qquad (3.103)$$

where
  k_{vs}, C_G is the valve constants
  ϕ is the valve orifice characteristic

Valve characteristics:

Two types are common, which are as follows:
  1. Linear characteristic

$$\phi_{lin} = \frac{k_{v0}}{k_{vs}} + 0.96\frac{x}{x_{max}} \qquad (3.104)$$

  2. Equal percentage characteristic

$$\phi_{e-p} = \frac{k_{vo}}{k_{vs}}e^{3.22}\frac{x}{x_{max}} \qquad (3.105)$$

  where
    k_{vo} is the valve constant
    x is the valve position
    x_{max} is the maximum valve of x

For simulation purposes, the equal-percentage characteristic will be used.

Drive motor characteristic:

  1. Linear characteristic of the valve drive motor.
  2. Set duration of 50 s between fully closed and fully opened states.

These assumptions have to be cross-checked and eventually adopted to the real behavior.

The actual position of the valve is compared with the setpoint given by the controller. If the difference between the two is less than a constant value ε, the valve position is not changed, so that the intense vibration of the valve motor between the opening and closing of the valve is avoided. One of the following expressions is used:

$$\left.\begin{array}{l} |x_{cont} - x| < \varepsilon \rightarrow dx/dt = 0 \\[2mm] x_{cont} - x > \varepsilon \rightarrow dx/dt = 0.02 \\[2mm] x_{cont} - x < \varepsilon \rightarrow dx/dt = -0.02 \end{array}\right\} \qquad (3.106)$$

where x_{cont} is the controller setpoint.

The flows in fluid streams in certain sections of an MSF plant are controlled. The valves controlling the following important streams are considered here:

1. LP steam flow
2. Blowdown flow
3. Distillate flow
4. Brine recirculation flow
5. Makeup flow seawater reject flow

The valves controlling these flows are all of equal percentage type. Such valves are deliberately chosen, because, despite their inherent nonlinear behavior, they tend to attain linear characteristics under installed conditions.

3.5.2 Controller Model

The control system in existing plants is based on the proved conventional single-loop concept with PI or PID:

$$m = K_P \left[e + \frac{1}{T_i} \int e\, dt + D \frac{de}{dt} \right] \qquad (3.107)$$

where
  m is the control variable
  e is the control error ($e = r - y$), which is the difference between setpoint r and measured value y

The control variable is thus a sum of three terms: P (proportional to the error), the I (which is proportional to the integral of the error), and the D term (which is proportional to the derivative of the error) (Figure 3.8).

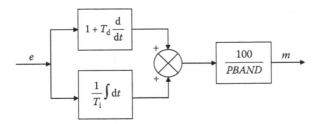

FIGURE 3.8
PID control loop.

The controller parameters are proportional to gain k_p, integral time T_i, and derivatives time T_d.

In general, each closed control loop involves variables that can be classified into three major categories: measured, manipulated, controlled. (These variables are flows, temperatures, pressure, valve position, etc.)

3.5.2.1 Other Subsystems, Correlations, and Interconnections

To complete the mathematical description of an operating MSF plant, models related to the various control systems are to be added to the subprocess models derived earlier. These control loops are equipped with PI, PID controllers, actuators, and valves in a standard structure to regulate process variables such as low-pressure steam temperature and flow, brine heater condensate level, brine level in the last stage, brine recirculation flow, makeup flow, seawater supply, and so on.

The complete MSF plant flowsheet can now be prepared with the descriptions of the subsystems and components. The PI, PID controller settings are taken from the actual plant. The correlations for the physical and thermodynamic properties are given in Appendix 3.A as provided in a set of subroutines. With the appropriate description of the interconnections among the various subsystem models, the dynamic model for a typical MSF plant containing 18-flash stages (15 recovery and 3 reject) is formulated in a set of equations that mathematically describe the process, in the next chapter.

3.6 Conclusion

Dynamic modeling of an MSF desalination plant is a complex exercise with various facets as described in this chapter. Appropriate importance and weight should be attached to each of these facets in order to obtain meaningful results from simulations. Moreover, when such a model is essentially a DAE model, one has to be careful to avoid higher index problems in its solution.

Auxiliary equations for calculating various properties for supporting model equations are presented in Appendix 3.A.

3.A Appendix: Important Correlations

3.A.1 Physical and Thermodynamic Properties

Specific heat capacity of water (Helal et al., 1986):

$$C_w = (1.001183 - 6.1666652 \times 10^{-5}T + 1.3999989 \times 10^{-7}T^2$$
$$+ 1.3333336 \times 10^{-9}T^3) \times 4.184 \qquad (3.A.1)$$

where
  T is in °F
  C_w is in kJ/(kg K)

Specific enthalpy of saturated water:

Equation 3.A.1 is integrated between the reference temperature of 32°F and the boiling temperature in °F to obtain

$$H_w = \left[\frac{\begin{array}{c} 1.0011833T - 3.0833326 \times 10^{-5}T^2 + 4.666663 \times 10^{-8}T^3 \\ + 3.333334 \times 10^{-10}T^4 - 31.92 \end{array}}{1.8 \times 4.184} \right] \qquad (3.A.2)$$

where
  T is in °F
  H_w is in kJ/kg

Density of pure water (Mothershed, 1966):

$$\rho_w = (62.707172 - 0.43955304 \times 10^{-2}T - 0.46076921 \times 10^{-4}T^2) \times 16.01846 \quad (3.A.3)$$

where
  T is in °F
  ρ_w is in kg/m³

Viscosity of pure water (Plant Vendor):

$$\mu_w = \frac{1.0 \times 10^{-3}}{\exp[3.244(T-20)/(m+109)]} \qquad (3.A.4)$$

where
  T is in °C
  μ_w is in N s/m²

Thermal conductivity of water (Plant Vendor):

$$K_w = 5.756 \times 10^{-4} + 1.526 \times 10^{-6} T - 5.81 \times 10^{-9} T^2 \tag{3.A.5}$$

where
  T is in °C
  K_w is in kW/(m°C)

Specific heat capacities of brine (Plant Vendor):

$$C_B = 4.185 - 5.381 \times 10^{-3} C + 6.26 \times 10^{-6} C^2 - (3.055 \times 10^{-5} + 2.774 \times 10^{-6} C$$
$$- 4.318 \times 10^{-8} C^2) T + (8.844 \times 10^{-7} + 6.527 \times 10^{-8} C - 4.003 \times 10^{-10} C^2) T^2 \tag{3.A.6}$$

where C, T, and C_B are in g/L, °C, and kJ/(kg °C), respectively.

Specific heat capacities of brine (Bromley et al., 1970):

$$C_B = [1.0 - C \times (0.011311 - 0.00001146 T)] C_{p,w} \tag{3.A.7}$$

where

$$C_{p,w} = 1.0011833 - 6.1666652 \times 10^{-5} T + 1.3999989 \times 10^{-7} T^2 + 1.3333336 \times 10^{-9} T^3$$

C, T, and C_B are in concentration in percentage, °F, and Btu/(lb°F), respectively.

Specific enthalpy of brine (Plant Vendor):

$$H_B = (4.185 - 5.381 \times 10^{-3} C + 6.26 \times 10^{-6} C^2) T$$
$$- (3.055 \times 10^{-5} + 2.774 \times 10^{-6} C - 4.318 \times 10^{-8} C^2) T^2 / 2$$
$$+ (8.844 \times 10^{-7} + 6.527 \times 10^{-8} C - 4.003 \times 10^{-10} C^2) T^3 / 3 \tag{3.A.8}$$

where C, T, and H_B are in g/L, °C, and kJ/kg, respectively.

Density of brine (Hoemig, 1978):

$$\rho_B = 0.5A_0 + A_1 Y + A_2(2Y^2 - 1) + A_3(4Y^3 - 3Y) \qquad (3.A.9)$$

where
$Y = (2T - 200)/160$
$\sigma = (2S - 150)/150$
$A_0 = 2.016110 + 0.115313\sigma + 0.000326\ (2\sigma^2 - 1)$
$A_1 = -0.0541 + 0.001571\sigma - 0.000423\ (2\sigma^2 - 1)$
$A_2 = -0.006124 + 0.00174\sigma - 0.000009\ (2\sigma^2 - 1)$
$A_3 = 0.000346 + 0.000087\sigma - 0.000053\ (2\sigma^2 - 1)$
C is the brine concentration (g/g)
S is the brine concentration (g/kg)
T is the temperature (°C)
C_B is the brine density (kg/L)

Viscosity of brine (Plant Vendor):

$$\mu_B = 1.002 \times 10^{-3} + 1.55 \times 10^{-6} + 8.5 \times 10^{-9} C^2 / \exp[3.244(T - 20)/(T + 109)] \quad (3.A.10)$$

where
T is in °C
μ_B is in N s/m²

Thermal conductivity of brine (Plant Vendor):

$$K_B = 5.756 \times 10^{-4} - 3.464 \times 10^{-5} C + 7.286 \times 0.06 C^2$$

$$+ (1.526 + 0.4662C - 0.2268C^2 - 0.02867C^3) \times 10^{-6} T$$

$$- (5.81 + 20.55C - 9.916C^2 + 1.464C^3) \times 10^{-9} T^2 \qquad (3.A.11)$$

where C, T, and K_B are in mass fraction, °C, and kW/(m°C), respectively.

Saturation pressure of steam (Griffin and Keller, 1965):

$$\log_{10}(P/218.167) = -(647.27 - T)/T\ [3.2437814 + 5.86826 \times 10^{-3}$$

$$(647.27 - T) + 1.1702379 \times 10^{-8}(647.27 - T)^3]/$$

$$\left\{1 + 2.1878462 \times 10^{-3}(647.27 - T)\right\} \qquad (3.A.12)$$

where
T is in K
P is in atmospheres

Vapor pressure of brine (Hoemig, 1978):

$$P_v = P_w (1 - 0.537 C) \tag{3.A.13}$$

where

$$\ln \frac{P_w}{P_c} = \frac{T_c}{T} \sum_{i=1}^{8} b_i \left[1 - \frac{T}{T_c} \right]^{(i+1)/2}$$

P_v is the vapor pressure of brine (bar)
P_w is the vapor pressure of pure water at the same temperature (bar)
C is the concentration (g/g)
P_c is the critical pressure (220.93 bar)
T_c is the critical temperature (647.25 K)
b_i = [−7.8889166, 2.5514255, −6.7161690, 33.239495, −105.38479, 174.35319, −148.39348, 48.631602]

Latent heat of vaporization of water (*Plant Vendor*):

$$l = 2495 - 2.132T - 2.632 \times 10^{-3} T^2 \tag{3.A.14}$$

where
T is in °C
l is in kJ/kg

Specific enthalpy of saturated steam (*Plant Vendor*):

$$H_v = 2499.15 + 1.955T - 1.927 \times 10^{-3} T^2 \tag{3.A.15}$$

where
T is in °C
H_v is in kJ/kg

Tube side heat transfer coefficient:

$$N_u = 0.027 Re^{0.8} Pr^{1/3} \tag{3.A.16}$$

Condensation coefficient on tubes (Plant Vendor):

$$h_o = 1.64327 \left[\frac{K_w^3 \rho_w^2}{(\mu_w w_F D_o) A_t} \right]^{1/3} (n)^{-(1/6)} \tag{3.A.17}$$

where
   K_w is the water thermal conductivity (kW/(m°C))
   ρ_w is the density of water (kg/m³)
   w_F is the condensate flowrate (kg/s)
   μ_w is the viscosity of water (N s/m²)
   D_o is the tube outside diameter (m)
   A_t is the total heat transfer area based on outside diameter of tube (m²)
   n is the number of tubes
   h_o is in kW/(m²°C)

Overall heat transfer coefficient:

$$U = \frac{1}{\left[\dfrac{1}{h_i} \times \dfrac{D_o}{D_i} + FF + \dfrac{D_o}{2K_m} \ln \dfrac{D_o}{D_i} + \dfrac{1}{h_c} \right]} \tag{3.A.18}$$

Overall heat transfer coefficient (Griffin and Keller, 1965):

$$U = \frac{1}{(z+y)} \tag{3.A.19}$$

After inclusion of fouling factor (*FF*), the coefficient becomes

$$U_o = \frac{U}{(1 + U \times FF)} \tag{3.A.20}$$

where

$$y = \frac{(V_j \times ID_j)^{0.2}}{\left[(160 + 1.92 T_{F,j}) V_j \right]}$$

$$z = 0.1024768 \times 10^{-2} - 0.7473939 \times 10^{-5} T_{D,j}$$

$$+ 0.999077 \times 10^{-7} T_{D,j}^2 - 0.430046 \times 10^{-9} T_{D,j}^3 + 0.6206744 \times 10^{-12} T_{D,j}^4$$

y is the brine-side film resistance
z is the sum of vapor side resistances
$T_{D,j}$ is the saturation temperature (°F)
$T_{F,j}$ is the brine temperature at exit (°F)
V_j is the linear velocity of brine (ft/s)
U and U_o are in Btu/(h ft²°F)

Boiling point elevation (Friedrich and Hafford, 1971):

$$BPE = [565.757/T - 9.81559 + 1.54739 \ln T - (337.178/T - 6.41981$$

$$+ 0.922753 \ln T) \times C + (32.681/T - 0.55368 + 0.079022 \ln T) \times C^2]$$

$$\{C/(266919.6/T^2 - 379.669/T + 0.334169)\} \tag{3.A.21}$$

where

$$\bar{C} = \frac{19.819 \times C}{(1-C)}$$

where C, T, and BPE are in mass fraction, K and °C, respectively.

3.A.2 Other Correlations

Nonequilibrium allowance (Helal et al., 1986):

$$\delta = 352 \times H_j^{1.1} \left(\Delta T_{B,j}\right)^{-0.25} \left(w_j \times 10^{-3}\right)^{0.5} \left(T_{D,j}\right)^{-25} \tag{3.A.22}$$

where
  j is the index for stage
  H is the brine level (in.)
  ΔT_B is the flashdown (°F)
  w is the chamber load (lb/(h ft width))
  T_D is the saturation temperature (°F)
  δ is the nonequilibrium allowance (°F)

Temperature loss across demister and condenser tubes

$$\Delta T = \frac{[\exp\ (1.885 - 0.02063 T_{D1})]}{1.8} \tag{3.A.23}$$

where $T_{D1} = T_D \times 1.8 + 32$ and T_D and ΔT are in °C.

Nomenclature

F_D	Distillate flowrate (kg/h)
F_I	Noncondensable flowrate (kg/h)
F	Cooling brine flowrate (kg/h)
g	Acceleration due to gravity (m/s²)
H	Specific enthalpy (kcal/kg¹)
h	Film heat transfer coefficient (kcal/h m²°C)

h_o	Orifice height (m)
h_W	Height of the kick-plate (m)
K_{DEM}	Demister constant
L	Brine level at vena contracta, tube length (m)
L_1	Brine level at the exit of upstream stage (m)
L_2	Brine level at the kick-plate (m)
L_3	Brine level at the exit of the stage (m)
$\bar{L}$	Brine level (m)
L_v	Jet height at vena contracta (m)
M	Holdup (kg)
P	Pressure (bar or atm)
P_1	Pressure in upstream stage (bar)
P_2	Stage pressure (bar)
Q	Rate of heat transfer (kcal/h)
R	Recycle flowrate (kg/h) Tube radius (m)
T	Temperature (°C)
T^*	Standard temperature (°C)
$\bar{T}$	Average temperature (°C)
t	Time variable (s)
U	Overall heat transfer coefficient (kcal/h m²°C)
u	Flow velocity (m/s)
u_1	Brine velocity approaching orifice (m/s)
u_2	Brine velocity just upstream of kick-plate (m/s)
u_v	Velocity of jet at vena contracta (m/s)
V	Vapor flowrate (kg/h)
W	Steam or condensate flowrate (kg/h)
w_o	Orifice width (m)
w_s	Stage width (m)
X_1	Hydraulic jump length (m)
X	Mass/mole fraction in liquid, condensate
Y	Mole fraction in vapor

Greek Symbols

ρ	Liquid density (kg/m³)
α	Activity
γ	Activity coefficient in liquid phase
φ	Activity coefficient in vapor phase

Subscripts

B	Brine
c	Condensate
D	Distillate
H	Brine heater
I	Noncondensable
IM	Inerts, mole fraction

In	Input
m	Makeup
ms	Motive steam
NEA	Nonequilibrium allowance
out	Output
s	Steam
sat	Saturated
sea	Seawater
T	Tubes
V	Vapor
w	Wall, water
WB	Water box

References

Abdelmessih, A.H. and I.C. Hsu (1976), The effect of hydraulic jump on the performance of a single-stage flash evaporator, *Desalination*, **19**, 65–74.

Al-Shayji, K.A. and Y.A. Liu (2002), Predictive modeling of large-scale commercial water desalination plants, data-based neural network and model-based process simulation, *Ind. Eng. Chem. Res.*, **41**(25), 6460–6474.

Aspen Tech (1991), *SPEEDUP User Manual*, Aspen Tech, Cambridge, MA.

Ball, S.J. (1986), Control of two-phase evaporating flows, *Desalination*, **59**, 199–217.

Bar, M. and M. Zeitz (1990), A knowledge-based flowsheet-oriented user interface for a dynamic process simulator, *Comput. Chem. Eng.*, **14**(11), 1275–1280.

Bromley, L.A., A.E. Diamond, E. Salam, and D.G. Wilkins (1970), Heat capacities and enthalpies of sea salt solutions to 200°C, *Journal of Chemical Engineering. Data*, **15**, 246.

Chow, V.T. (1959), *Open Channel Hydraulics*, McGraw-Hill Koga Kusha Ltd, Tokyo, Japan.

Ciba–Geigy (1978), Non-condensable gases and the venting of seawater evaporators, Bulletin DB 2.2.

Delene, J.G. and S.J. Ball (1971), A digital computer code for simulating large multistage flash evaporator desalting plant dynamics, Oak Ridge National Laboratory, Oak Ridge, TN, Report ORNLTM-2933.

Dearth, J.D., T.M. McAuley, and A.M. Sheikh (1970), *Single-Phase Flow Regimes in a Multistage Flash Evaporator Stage*, ORNL MIT-94, 35pp.

Deutsche Babcock AG Report Vol. I (1996), Annex C, Report by Wangnick Consulting GMBH, Gnarrenburg, Germany.

Deutsche Babcock AG Report Vol. VII, Chapter 9, Report by Wangnick Consulting GMBH, Gnarrenburg, Germany.

Deutsche Babcock AG Report Vol. II (1996), Annex M, Report by Wangnick Consulting GMBH, Gnarrenburg, Germany.

Dooly, R. and J. Glater (1972), Alkaline scale formation in boiling seawater brine, *Desalination*, **11**, 1–17.

El Hisham, D. (1981), Thermodynamic and hydrodynamic behavior of orifices of seawater desalination plants, PhD thesis, Glasgow University, Glasgow, UK.

Emad, A. (2002a), Understanding the operation of industrial MSF plants Part II: Optimization and dynamic analysis, *Desalination*, **143**, 73–91.

Emad, A. (2002b), Understanding the operation of industrial MSF plants Part I: Stability and steady-state analysis, *Desalination*, **143**, 53–72.

Friedrich, R.O. and A.J. Hafford (1971), Rep. ORNL-TM-3489. Oak Ridge National Laboratory.

Fukuri, A., K. Hamanaka, M. Tatsumoto, and A.S. Inshara (1985), Automatic control system of MSF process (ASCODES), *Desalination*, **55**, 77–89.

Gambier, A. and B. Essameddin (2004), Dynamic modeling of MSF plants for automatic control and simulation purposes: A survey, *Desalination*, **166**, 191–204.

Gerd Posch (1977), Investigation of pressure losses in the evaporator of large MSF plants, PhD Thesis, Glasgow University, Glasgow, UK.

Glater, J., J.L. York, and K.S. Campbell (1980), Scale formation and prevention, in *Principles of Desalination* (eds. K.S. Spiegler and A.D.K. Laird), Academic Press, New York.

Glueck, A.R. and R.W. Bradshaw (1970), A mathematical model for a multistage flash distillation plant, in *Proceedings of the Third International Symposium on Fresh Water from the Sea*, Athens, Greece, Vol. 1, pp. 95–108.

Griffin, W.L. and R.M. Keller (1965), Rep. ORNL-TM-1299. Oak Ridge National Laboratory.

Hamer, P. et al. (1961), *Industrial Water Treatment Practice*, Butterworth, London, UK.

Hancke, K. (1994), *Wasseraufbereitungschemie und Chemische Verfahrenstechnik*, VDI Verlag, Düsseldorf.

Heitmann, H.G. (1990), *Saline Water Processing*, VCH Verlagsgessellschaft, Weinheim, Germany.

Helal, A.M., M.S. Medani, M.A. Soliman, and J.R. Flower (1986), A tridiagonal matrix model for flash desalination plants, *Computers and Chemical Engineering*, **10**, 327–342.

Hillal, M.M. and M.A. Marwan (1985), Design equations for a new setup of sluice gates to stabilize interstage brine flow, *Desalination*, **55**, 139–144.

Hiller, H. (1952), *Proceedings of the Institute of Mechanical Engineers*, London, UK, p. 295.

Holl, P., W. Marquardt, and E.D. Gilles (1988), Diva—A powerful tool for dynamic process simulation, *Comput. Chem. Eng.*, **12**(5), 421–426.

Hoemig, H.E. (1978), *Fitchner Handbook on Seawater and Seawater Distillation*, Vulkan Verlag, Essen, Germany.

Husain, A., K.V. Reddy, and A. Woldai (1994), Modeling, simulation and optimization of an MSF desalination plant, in *Eurotherm Seminar*, Thessaloniki, Greece.

Husain, A., A. Woldai, A. Hassan, D.M.K. Al-Gobaisi, A.A. Radif, and C. Sommariva (1992), Modeling, simulation, optimization and control of MSF desalination plants, part I. Modeling and simulation, *Desalination*, **92**, 21–41.

Husain, A., A. Woldai, A.A. Radif, A. Kesou, R. Borsani, H. Sultan, and P.B. Deshpandey (1993), Modeling and simulation of an MSF desalination plant, *Desalination*, **97**, 555–586.

Ishihara, T. and Ida, T. (1951), Supplemented formulas for rectangular weirs without end-contractions, *Proceedings of the First Japan National Congress for Applied Mechanics*, Japan National Committee for Theoretical and Applied Mechanics, Science Council of Japan, Tokyo.

Kamali, R.K., A. Abbassi, and S.A. Sadough Vanini (2009), A simulation model and parametric study of MED-TVC process, *Desalination*, **235**, 340–351.

Khan, A.H. (1986), *Desalination Processes and Multistage Flash Distillation Practice*, Elsevier, Amsterdam, Netherlands.

Kishi, M.K., K. Matsumoto, Y. Takerchi, and K. Hattori (1985), Development of flow model for weir type orifice, *Desalination*, **55**, 481–492.

Kishi, M.K., Y. Mochizuki, M. Matsubayashi, and K. Hattori (1987), Development of flashing flow models for box type orifice, *Desalination*, **65**, 57–62.

Kroner, A., P. Holl, W. Marquardt, and E.D. Gills (1990), Diva—An open architecture for dynamic simulation, *Comput. Chem. Eng.*, **14**(11), 1289–1295.

Langlier, W.F., D.H. Caldwell, and W.B. Lawrence, Institute of Engineering Research, University of California, final report. Contract W-44-009 Eng.-499, Engineer Research and Development Laboratories, Ft. Belvoir, August 15, 1950.

Lefkopolous, A. and M.A. Stadtherr (1993), Index analysis of unsteady-state chemical process systems-I. An algorithm for problem formulation, *Comput. Chem. Eng.*, **17**(4), 399–413.

Lior, N. (1986), Formulas for calculating the approach to equilibrium in open channel flash evaporators for saline water, *Desalination*, **60**, 223–249.

Maniar, V.M. and P.B. Deshpandey (1996), Advanced controls for MSF desalination plants, *J. Process Control*, **6**(1), 49–66.

Marquardt, W. (1995), Trends in computer-aided process modeling, in *PSE '94*, Kyongju, Korea.

Marquardt, W. (1996a), Report 1-A: Towards a comprehensive MSF model; model validation for UANE 4-6 MSF plants, Wangnick Consulting GMBH, Gnarrenburg, Germany.

Marquardt, W. (1996b), Modeling and simulation of noncondensable gases in MSF desalination, Report 2-A, Wangnick Consulting GMBH, Gnarrenburg, Germany.

Marquardt, W. (1996c), Rigorous dynamic modeling and simulation of electrolyte systems, Report 2-A, Wangnick Consulting GMBH, Gnarrenburg, Germany.

Mohamed, A. and A. Abdulnaser (2008), Steady state simulation of MSF desalination plant, ISESCO Science and Technology Vision, Rabat, Morocco, Vol. 4, No. 5, pp. 30–33

Mothershed, C.T. (1966), Rep. ORNL-TM-1560. Oak Ridge National Laboratory.

Pantelides, C.C., D. Gritals, K.R. Morisons, and R.W.H. Sargent (1988), The mathematical modeling of transient systems using differential-algebraic equations, *Comput. Chem. Eng.*, **12**(5), 449–454.

Perry, R.H. and D. Green (1984), *Perry's Chemical Engineers Handbook*, 6th edn., McGraw Hill, New York.

Peter, P. and D. Bijan (2003), Integrated thermal power and desalination plant optimization, General Electric Energy Services, Optimization Software, PowerGen Middle East, October 2003, Paper No. 110.

Reddy, K.V., A. Husain, A. Woldai, and M.K.A. Darwish (1995a), Dynamic modeling of the MSF desalination process, in *Proceedings of IDA World Congress on Desalination and Water Sciences*, Abu Dhabi, United Arab Emirates, Vol. IV, pp. 227–242.

Reddy, K.V., A. Husain, A. Woldai, S.M. Nabi, and A. Kurdali (1995b), Holdup and interstage orifice flow model for an MSF desalination plant, in *Proceedings of IDA World Congress on Desalination and Water Sciences*, Abu Dhabi, United Arab Emirates, Vol. IV, pp. 323–340.

Rimawi, M.A., H.M. Eltouney, and G.S. Aly (1989), Transient model of multistage flash desalination, *Desalination*, **74**, 327–338.

Seifert, A. (1988), Das Intergas Problem und Verlusteffetke in Entspannungverdampfern für die Meerwasserentsalzung, Dissertation, University of Bremen, Bremen, Germany.

Seifert, A. and K. Genthner (1991), A model for stagewise calculation of non-condensable gases in multi-stage evaporators, *Desalination*, **81**, 333–347.

Shams, E.L., A.M. Din, and R.A. Mohammed (1989a), On the thermal stability of the HCO_3^- and CO_3^{2-} ions in aqueous solutions, *Desalination*, **69**, 241–249.

Shams, E.L. A.M. Din, and R.A. Mohammed (1989b), The problem of alkaline scale formation from a study on Arabian Gulfwater, *Desalination*, **71**, 313–324.

Smith, G.J. and W. Morton (1988), Dynamic simulation using an equation-oriented flowsheeting package, *Comput. Chem. Eng.* **12**(5), 469–473.

Subramanya, K. (1982), *Flow in Open Channels*, Tata McGraw Hill Co, New Delhi, India.

Ulrich, J. (1977), Dynamic behaviors of MSF plants for seawater desalination, PhD Dissertation, University of Hannover, Hannover, Germany.

Unger, J., A. Kroner, and W. Marquardt (1995), Structural analysis of differential-algebraic equation systems-theory and applications, *Comput. Chem. Eng.*, **19**(8), 867–882.

Unger, J. and W. Marquardt (1991), Structural analysis of differential-algebraic equation systems, in *Computer-Aided Process Engineering Proceedings of COPE 91* (eds. L. Pingjaner and A. Espuna), Elsevier Science Publications, Amsterdam, Netherlands, pp. 241–246.

Villemonte, J.R. (December 25, 1947), Submerged-weir discharge studies, *Eng. News-Record*, **139**, 54.

Watson Desalination Consultants (1979), *Technology Review and Handbook: High Temperature Scale Inhibitors for Sea Water Distillation. A Multi-Client Study*, Watson Desalination Consultants, Manassas, VA.

4

Data Reconciliation and Model Validation with Experimental Data

The model obtained on the basis of physical laws needs to be validated using experimental data. Practically, it is very important to render the model fit for further use. At first, the data obtained from plant measurements need to be filtered if there is noise. For this, there are many filtering techniques. However, in our case, the process is sufficiently slow to reject all high-frequency noises. Low-frequency noise needs to be tackled in a different way. This is known as data reconciliation. After correcting the measured data by this method, one can proceed further with the use of the data. We, therefore, will not consider the filtering methods but will first illustrate the process of data reconciliation that is relevant to this situation.

4.1 Data Reconciliation

Let M, B, and D denote the steady-state measurements of makeup flow, blowdown flow, and distillate flow, all rates, respectively, in an MSF desalination plant. These measured values are not directly useful as they are usually corrupted by measurement errors. As they are, these measurements do not satisfy the important condition of material balance, namely, $M \neq B + D$. In order to render them worthy of further use, we resort to data reconciliation to yield the corresponding values $\hat{M} = \hat{B} + \hat{D}$. This can be formulated as simple linear static data reconciliation problem.

Minimize $F = (M - \hat{M})^2 + (B - \hat{B})^2 + (D - \hat{D})^2$ with respect to $\hat{M}$, $\hat{B}$, and $\hat{D}$ subject to the condition $\hat{M} = \hat{B} + \hat{D}$.

This constrained optimization problem can be solved by the method of Lagrange multipliers as an unconstrained one. That is, with the Lagrange multiplier l, we define

$$\hat{F} = F + \lambda (\hat{M} - \hat{B} - \hat{D})$$

and minimize it with respect to $\hat{M}$, $\hat{B}$, $\hat{D}$, and l.

The conditions for minimum are

$$\frac{\partial \hat{F}}{\partial \hat{M}} = M - \hat{M} - \frac{\lambda}{2} = 0$$

$$\frac{\partial \hat{F}}{\partial \hat{B}} = B - \hat{B} - \frac{\lambda}{2} = 0$$

$$\frac{\partial \hat{F}}{\partial \hat{D}} = D - \hat{D} - \frac{\lambda}{2} = 0$$

$$\frac{\partial \hat{F}}{\partial \lambda} = \hat{M} - \hat{B} - \hat{D} = 0$$

In one situation in an MSF plant, the following measurements were made: $M = 5360$ t/h, $B = 4025$ t/h, and $D = 1188$ t/h. Note that $M \neq B + D$.

Inserting these values in the aforementioned four equations, we get the solution as

$$\hat{M} = 5311 \, t/h \quad \hat{B} = 4074 \, t/h \quad \hat{D} = 1237 \, t/h \quad \text{and} \quad \lambda = 98$$

Note that $\hat{M} = \hat{B} + \hat{D}$.

The aforementioned solution can be generalized to the case of measurements with several constraints in the optimization problem. This is applicable to temperature measurements of several stages in an MSF plant with group-wise equations for thermal balance.

4.2 General Linear Static Data Reconciliation Problem

Consider the set of temperature measurements $\theta_i^*, i = 1, 2, \ldots, m$ and k thermal balance constraints. The cost function to be minimized is defined by

$$\hat{F}(\theta) = \frac{1}{2} \sum_{i=1}^{m} \left(\theta_i - \theta_i^* \right)^2 + \sum_{j=1}^{k} \lambda_j \left(\sum_{i=1}^{m} a_{j,i} \theta_i - q_j \right)$$

which can be compactly written in matrix form as

$$\hat{F}(\theta) = \frac{1}{2}(\theta - \theta^*)^T(\theta - \theta^*) + \lambda^T(A\theta - q)$$

where

θ^* is the m-vector of measurements $= \begin{bmatrix} \theta_1^* & \theta_2^* & \cdots & \theta_m^* \end{bmatrix}^T$

$\boldsymbol{\theta}$ is the m-vector of estimated measurements $= \begin{bmatrix} \theta_1 & \theta_2 & \cdots & \theta_m \end{bmatrix}^T$

$\mathbf{q}$ is the k-vector of heat loss values $= \begin{bmatrix} q_1 & q_2 & \cdots & q_k \end{bmatrix}^T$

$\mathbf{A}$ is the $(k \times m)$-matrix of coefficients in the constraint equations

λ is the k-vector of the Lagrange multipliers

The conditions for minimum are obtained by equating the partial derivatives of $\hat{F}$ w.r.t. θ_i, $i = 1,2,\ldots,m$ and λ_j, $j = 1,2,\ldots,k$ as

$$\theta_i + \sum_{j=1}^{k} a_{j,i}\lambda_j = \theta_i^*, \quad i = 1,2,\ldots,m$$

and

$$\sum_{i=1}^{m} a_{j,i}\theta_i = q_j, \quad j = 1,2,\ldots,k$$

These conditions can be compactly written in matrix form as

$$\begin{bmatrix} \dfrac{\partial\hat{F}}{\partial\theta} \\[2ex] \dfrac{\partial\hat{F}}{\partial\lambda} \end{bmatrix} = \begin{bmatrix} \mathbf{I}_{m\times m} & \mathbf{A}^T \\ \mathbf{A} & 0 \end{bmatrix}\begin{bmatrix} \theta \\ \lambda \end{bmatrix} = \begin{bmatrix} \theta^* \\ q \end{bmatrix}$$

whose solution is

$$\begin{bmatrix} \theta \\ \lambda \end{bmatrix} = \begin{bmatrix} \mathbf{I}_{m\times m} & -\mathbf{A}^T(\mathbf{A}\mathbf{A}^T)^{-1}\mathbf{A} & \mathbf{A}^T(\mathbf{A}\mathbf{A}^T)^{-1} \\ & +(\mathbf{A}\mathbf{A}^T)^{-1}\mathbf{A} & -(\mathbf{A}\mathbf{A}^T)^{-1} \end{bmatrix}\begin{bmatrix} \theta^* \\ q \end{bmatrix}$$

which exists only if $(\mathbf{A}\mathbf{A}^T)^{-1}$ exists. See Cameron (1996) for more details of the data reconciliation problem.

4.3 Validation of the MSF Process Model

In the case of dynamic data with nonlinear constraints, the present formulation leads to the so-called nonlinear dynamic data reconciliation (Liebman et al., 1992) problem.

The MSF process model described in Chapter 3 is one which seems appropriate in form for simulation on the flowsheet system of SPEEDUP. Although the numerical values of many of the parameters of the model in terms of physical dimensions, constants, and correlations are inserted in appropriate places, there exist some key characteristics such as those of the control valves, which need to be ascertained. Fortunately, to aid in this process, we have measurement data on the important process variables that are available. These data, obtained from static and dynamic tests at a few operating points, are helpful in fixing the control characteristics of the valves. This chapter presents model validation using extensive static and dynamic measurement data from the plant for the temperature, flow, and valve position of the stages with corresponding flow.

4.3.1 Valve Models

The valve that controls the steam flow into the brine heater is of much importance in the MSF plant. This valve is modeled by the characteristic

$$F = C_v R^{m-1} \sqrt{\Delta P}$$

where F, C_v (=7), m and ΔP denote, respectively, the flow, valve coefficient, fractional position of the valve, and pressure drop. A rangeability factor of 10 in the equal percentage characteristic is found to adequately describe the valve behavior as shown in Figure 4.1.

Figure 4.2 shows the characteristic curve; top brine temperature (TBT) versus the LP steam flow of (the measured and model data).

4.3.2 Other Valves

Five other liquid service valves that are important in this context are considered. These are all also designed for equal percentage behavior. In the light of the fact that each of these valves operates in not too wide a flow range of its own, the installed characteristics of these valves are modeled in the form

$$F = F_{\max} R_{\text{eff}}^{m-1}$$

where F, $F_{\max}$, R_{eff}, and m are flow, flow at full opening, effective rangeability, and fractional valve position, respectively. This description is adequate for the purpose of control. Table 4.1 shows the model parameters in the installed characteristics of these liquid delivery valves.

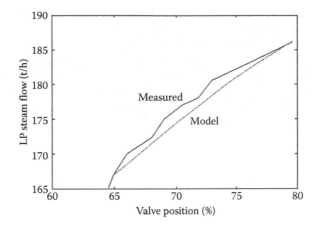

FIGURE 4.1
Characteristic curve for LP steam flow versus control valve position.

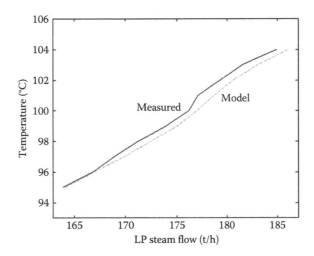

FIGURE 4.2
Characteristic curve for TBTs versus LP steam flow.

TABLE 4.1

Model Parameters in the Installed Characteristics
of Liquid Delivery Valves

Valve	F_{max} (t/min)	R_{eff}
Blowdown	120.2233	6.9338
Distillate flow	50.745	4.414
Brine recirculation	983.3833	9.4977
Makeup	221.4833	5.5927
Seawater reject	225.34	5.7345

Figure 4.3a through 4.3e shows the behavior of the valves modeled as mentioned with their actually observed characteristics.

With six of the most essential valves in the plant satisfactorily modeled as mentioned, we do not stress the need, although it is ideally desired, to tune the other parts of the process model (e.g., orifices, flash chamber dynamics, etc.) for the purpose of this investigation. This is justified, since the control design concepts we have in mind are expected to be tolerant to model inaccuracy and uncertainty to some extent.

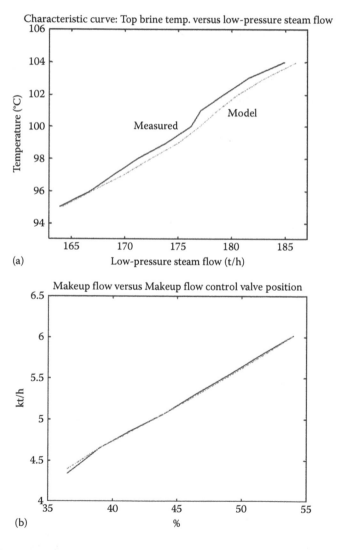

(a)

(b)

FIGURE 4.3
(a–e) Characteristics of liquid service valves (———, Measured; –·–·–·–, Estimated).   (*Continued*)

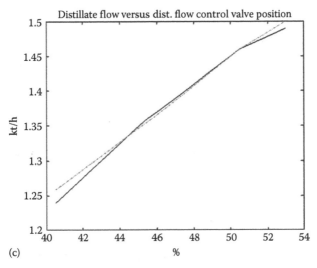

(c)

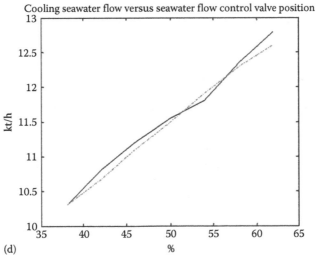

(d)

FIGURE 4.3 (*Continued*)
(a–e) Characteristics of liquid service valves (————, Measured; —·—·—·—, Estimated). (*Continued*)

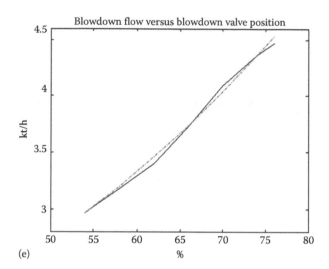

(e)

FIGURE 4.3 (*Continued*)
(a–e) Characteristics of liquid service valves (————, Measured; —·—·—·—, Estimated).

In fact, the behavior of the model is compared with that of the actual plant and found to be satisfactory. This is done by conducting static and dynamic simulations and comparing with the actual plant measurements.

4.4 Steady-State Simulation Results and Comparisons

The steady-state performance of an MSF plant, under operation in Abu Dhabi, consisting of 15 recovery and 3 rejection stages, has been simulated. Table 4.2 compares the results of simulation using the present model with the actual test data of the plant at a particular operating condition.

The results show the temperatures of the brine, distillate, and cooling tubes in 18 flash stages.

Tables 4.A.1 through 4.A.12 in the Appendix show the results of steady-state simulation in 12 different cases or operating conditions. These simulations have taken two important process variables, that is, the TBT to which the brine is heated in the brine heater, and the brine recycle rate. The maximum allowable TBT depends upon the scale inhibitors added to makeup feed. In these examples, Tables 4.A.1 through 4.A.12 are given with the recycle rates 11,500, 12,500, 13,500, and 14,420 t/h keeping the TBT constant to 95°C, 100°C, and 105°C at a time, respectively, without violating the

TABLE 4.2

Comparison of Summer Temperature Profile Results Obtained from Simulated (Sim) and Observed (Obs) in Steady State

	Simulation Result	Observed Result
Makeup flow	5520	5516 (t/h)
Blowdown flow	4383	4376 (t/h)
Product flow	1133	1140 (t/h)
Reject flow	8988	8983 (t/h)
TBT	90°C	90°C
Performance ratio	7.20	7.02 (kg/540 kcal)

Flash Stage No.	Brine (°C)		Distillate (°C)		Cooling Tube (°C)	
	Sim	Obs	Sim	Obs	Sim	Obs
1	86.95	87.03	85.88	85.93	80.71	80.70
2	83.97	84.11	82.90	83.03	77.73	77.79
3	81.00	81.20	79.93	80.12	74.76	74.89
4	78.06	78.23	76.97	77.15	71.81	71.93
5	75.13	75.28	74.04	74.18	68.88	68.98
6	72.22	72.35	71.12	71.24	65.96	66.06
7	69.32	69.44	68.21	68.32	63.06	63.16
8	66.45	66.56	65.33	65.41	60.18	60.28
9	63.60	63.71	62.46	62.54	57.33	57.43
10	60.78	60.89	59.62	59.67	54.50	54.62
11	57.98	58.09	56.79	56.87	51.70	51.82
12	55.23	55.33	53.99	54.06	48.94	49.07
13	52.51	52.62	51.23	51.29	46.21	46.35
14	49.83	49.95	48.50	48.56	43.53	43.69
15	47.20	47.33	45.81	45.88	40.88	41.07
16	45.03	45.16	43.61	43.68	39.29	39.24
17	42.91	43.02	41.44	41.53	37.21	37.13
18	40.88	40.88	39.51	39.45	35.00	35.00

constraints on flow velocity through the tubes and flashing brine levels, in order to achieve the desired production rate or performance efficiency. The seawater flowrate and its temperature are important in keeping the proper flashing temperature in the first and last rejection stages. The brine level in the last stage is controlled so as to maintain the proper levels in the preceding stages between one extreme of excessive submergence and the other a *blowthrough* condition. The results in Tables 4.A.1 through 4.A.12 show the temperatures of the brine, distillate, cooling tubes, and distillate production in each stage of the 18-stage flash.

The good agreement between the temperature profiles of the model and the actual plant should be viewed only as a partial success in the model validation. Steady-state simulations, with closed-loop control systems having PI controllers, reveal nothing about the dynamic behavior that is required for control system considerations. However, this is our first successful step in model verification.

4.5 Dynamic Simulation and Model Verification

Figures 4.4 and 4.5 show the transient profiles of brine temperature of the first and last stages. This is a typical result of an open-loop test carried out by giving a step change in the steam supply rate (reduction at 1000s and back to the original value at 10000s). The dynamic simulation is run under similar conditions, and its result is shown in Figures 4.4 and 4.5. Qualitatively, both the results sufficiently agree; however, the simulation result shows that the TBT profile settles down much more quickly than the same in the plant test.

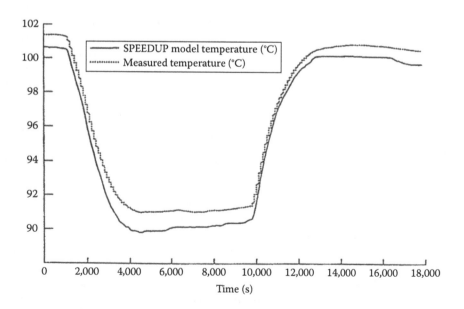

FIGURE 4.4
Transient profiles of the brine temperature at the first stage.

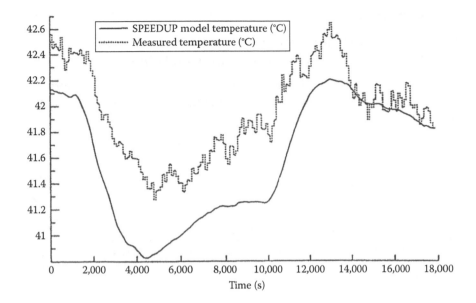

FIGURE 4.5
Transient profiles of the brine temperature at the last stage.

4.6 Summary and Discussion

Model validation strategies have been investigated. These are validations using measured plant data and the rigorous nonlinear model. The use of the rigorous model is highly computational demanding and requires a considerable implementation effort. The model is acceptable, in view of its satisfactory agreement in dynamic behavior with actual plant data for further consideration such as analysis and control system design.

The results shown in Figures 4.4 and 4.5 are particularly good and show a high degree of correlation between the model and the real system measurements. The author is not aware of any better correlation between model and real system measurements for this plant type within the literature.

The main difference between the model response and the real system response in both cases of Figures 4.4 and 4.5 is a constant offset. This constant disparity between the model response and the measured temperatures could be due to a small bias in the temperature measuring system or recorder. Note that both Figures 4.4 and 4.5 have a large false origin.

4.A Appendix: Steady-State Performance of the MSF Plant under Different Operating Conditions

The material contained within this appendix gives steady-state performance in terms of key variables at each of the 18 stages for each of the 12 different operating conditions.

	Recycle Flow (t/h)			
TBT (°C)	**14,420**	**13,500**	**12,500**	**11,500**
95	Case 1	Case 4	Case 7	Case 10
100	Case 2	Case 5	Case 8	Case 11
105	Case 3	Case 6	Case 9	Case 12

Note: Case 1 (Table 4.A.1); Case 2 (Table 4.A.2); Case 3 (Table 4.A.3); Case 4 (Table 4.A.4); Case 5 (Table 4.A.5); Case 6 (Table 4.A.6); Case 7 (Table 4.A.7); Case 8 (Table 4.A.8); Case 9 (Table 4.A.9); Case 10 (Table 4.A.10); Case 11 (Table 4.A.11); Case 12 (Table 4.A.12).

The following symbols are used in this appendix:

TF_IN   Temperature of the cooling brine entering a stage
D_OUT   Distillation outlet flowrate from a stage
TD_OUT   Temperature of the distillate leaving a stage
TB_OUT   Flash brine temperature leaving a stage
B_OUT   Brine outlet flowrate leaving a stage

TABLE 4.A.1

Case 1

Operating conditions:					
Steam flow	165.733	[t/hr]	Reject flow	6,297.182	[t/h]
Recycle flow	14,419.995	[t/hr]	Makeup flow	6,142.800	[t/h]
Blowdown flow	4,919.913	[t/h]	Top brine temperature	95.000	[°C]
Product flow	1,219.131	[t/h]	Seawater flow	12,439.982	[t/h]
Performance ratio	7.312	[kg/540 kcal]			

Stages	TF_IN	D_OUT	TD_OUT	B_OUT	TB_OUT
1	85.102	1.323	90.640	238.977	91.712
2	81.897	2.628	87.432	237.666	88.502
3	78.702	3.915	84.242	236.377	85.313
4	75.524	5.183	81.056	235.107	82.137
5	72.360	6.432	77.894	233.856	78.979
6	69.214	7.662	74.744	232.625	75.839
7	66.087	8.873	71.615	231.412	72.716
8	62.981	10.063	68.502	230.220	69.616
9	59.900	11.234	65.412	229.048	66.541
10	56.845	12.383	62.344	227.897	63.491
11	53.822	13.509	59.296	226.769	60.475
12	50.837	14.612	56.279	225.666	57.497
13	47.893	15.689	53.296	224.587	54.560
14	44.994	16.740	50.348	223.535	51.670
15	42.134	17.769	47.452	222.505	48.820
16	40.651	18.554	45.282	221.719	46.625
17	38.039	19.363	42.908	220.908	44.348
18	35.000	20.319	40.751	81.999	42.134

TABLE 4.A.2

Case 2

Operating conditions:

Steam flow	1,180.167	[t/h]	Reject flow	6,297.182	[t/h]
Recycle flow	14,419.995	[t/h]	Makeup flow	6,142.800	[t/h]
Blowdown flow	4,817.894	[t/h]	Top brine temperature	100.000	[°C]
Product flow	1,320.581	[t/h]	Seawater flow	12,439.982	[t/h]
Performance ratio	7.363	[kg/540 kcal]			

Stages	TF_IN	D_OUT	TD_OUT	B_OUT	TB_OUT
1	89.372	1.441	95.326	238.852	96.430
2	85.853	2.861	91.849	237.425	92.949
3	82.386	4.262	88.389	236.023	89.489
4	78.935	5.641	84.934	234.640	86.041
5	75.497	7.000	81.500	233.280	82.610
6	72.078	8.3372	78.079	231.941	79.197
7	68.679	9.652	74.679	230.624	75.802
8	65.302	10.944	71.297	229.330	72.432
9	61.950	12.214	67.939	228.059	69.089
10	58.627	13.460	64.604	226.811	65.772
11	55.340	14.679	61.293	225.590	62.492
12	52.094	15.872	58.014	224.396	59.254
13	48.893	17.036	54.7726	223.230	56.062
14	45.742	18.170	51.570	222.095	52.922
15	42.635	19.279	48.423	220.985	49.827
16	41.088	20.120	46.092	220.143	47.45
17	38.277	20.984	43.532	219.277	45.003
18	35.000	22.010	41.224	80.298	42.635

TABLE 4.A.3

Case 3

Operating conditions:					
Steam flow	195.016	[t/h]	Reject flow	6297.180	[t/h]
Recycle flow	14,419.980	[t/h]	Makeup flow	6142.800	[t/h]
Blowdown flow	4,716.280	[t/h]	Top brine temperature	105.000	[°C]
Product flow	1,412.550	[t/hr]	Seawater flow	12439.980	[t/h]
Performance ratio	7.402	[kg/540 kcal]			

Stages	TF_IN	D_OUT	TD_OUT	B_OUT	TB_OUT
1	93.554	1.557	100.011	238.728	101.152
2	89.814	3.092	96.268	237.185	97.403
3	86.079	4.607	92.543	235.668	93.675
4	82.356	6.097	88.819	234.174	89.956
5	78.646	7.565	85.116	233.704	86.253
6	74.954	9.0092	81.424	231.258	82.567
7	71.282	10.429	77.754	229.836	78.901
8	67.633	11.823	74.102	228.440	75.260
9	64.010	13.192	70.476	227.071	71.648
10	60.418	14.534	66.874	225.727	68.063
11	56.865	15.846	63.297	224.412	64.518
12	53.356	17.128	59.756	223.129	61.018
13	49.896	18.379	56.254	221.877	57.568
14	46.492	19.596	52.796	222.658	54.176
15	43.137	20.783	49.398	219.470	50.835
16	41.526	21.680	46.909	218.571	48.287
17	38.515	22.598	44.157	217.651	45.658
18	35.000	23.693	41.697	78.605	43.137

TABLE 4.A.4

Case 4

Operating conditions:					
Steam flow	151.308	[t/h]	Reject flow	6,297.180	[t/h]
Recycle flow	13,500.000	[t/h]	Makeup flow	6,142.800	[t/h]
Blowdown flow	4,984.373	[t/h]	Top brine temperature	95.000	[°C]
Product flow	1,154.700	[t/h]	Seawater flow	12,439.980	[t/h]
Performance ratio	7.526	[kg/540 kcal]			

Stages	TF_IN	D_OUT	TD_OUT	B_OUT	TB_OUT
1	85.167	1.261	90.593	223.705	91.651
2	81.907	2.505	87.329	222.456	88.385
3	78.657	3.731	84.084	221.228	85.141
4	75.426	4.938	80.845	220.018	81.911
5	72.209	6.128	77.629	218.828	78.700
6	69.011	7.298	74.427	217.655	75.507
7	65.833	8.450	71.246	216.502	72.334
8	62.678	9.582	68.084	215.368	69.184
9	59.548	10.695	64.944	214.255	66.060
10	56.445	11.787	61.828	213.161	62.963
11	53.378	12.857	58.734	212.090	59.901
12	50.348	13.903	55.673	211.042	56.878
13	47.362	14.925	52.646	210.018	53.900
14	44.422	15.922	49.658	209.020	50.969
15	41.525	16.896	46.722	208.045	48.081
16	40.147	17.638	44.525	207.302	45.865
17	37.745	18.381	42.210	206.558	43.629
18	35.000	19.245	40.170	83.073	41.525

TABLE 4.A.5

Case 5

Operating conditions:

Steam flow	164.430	[t/h]	Reject flow	6,297.180	[t/h]
Recycle flow	13,500.000	[t/h]	Makeup flow	6,142.800	[t/h]
Blowdown flow	4,887.774	[t/h]	Top brine temperature	100.000	[°C]
Product flow	1,250.738	[t/h]	Seawater flow	12,439.980	[t/h]
Performance ratio	7.58	[kg/540 kcal]			

Stages	TF_IN	D_OUT	TD_OUT	B_OUT	TB_OUT
1	89.398	1.373	95.275	223.587	96.363
2	85.865	2.727	91.737	222.226	92.822
3	82.339	4.062	88.219	220.890	89.303
4	78.830	5.375	84.705	219.573	85.796
5	75.336	6.668	81.214	218.278	82.307
6	71.860	7.940	77.736	217.004	78.838
7	68.405	9.192	74.281	215.751	75.388
8	64.975	10.421	70.844	214.520	71.964
9	61.571	11.627	67.433	213.313	68.568
10	58.197	12.811	64.046	212.127	65.200
11	54.861	13.969	60.685	210.968	61.871
12	51.567	15.101	57.358	209.834	58.584
13	48.320	16.205	54.070	208.728	55.347
14	45.126	17.281	50.824	207.652	52.163
15	41.979	18.331	47.635	206.601	49.028
16	40.547	19.124	45.275	205.806	46.635
17	37.962	19.918	42.779	205.011	44.227
18	35.000	20.846	40.598	81.463	41.979

TABLE 4.A.6

Case 6

Operating conditions:

Steam flow	179.109 [t/h]		Reject flow	6,297.180 [t/h]
Recycle flow	13,500.000 [t/h]		Makeup flow	6,142.800 [t/h]
Blowdown flow	4,791.549 [t/h]		Top brine temperature	105.000 [°C]
Product flow	1,346.318 [t/h]		Seawater flow	12,439.980 [t/h]
Performance ratio	7.620 [kg/540 kcal]			

Stages	TF_IN	D_OUT	TD_OUT	B_OUT	TB_OUT
1	93.631	1.485	99.957	223.468	101.079
2	89.828	2.947	96.149	221.997	97.266
3	86.029	4.390	92.360	220.552	93.474
4	82.245	5.810	88.573	219.128	89.692
5	78.473	7.207	84.808	217.729	85.927
6	74.721	8.581	81.056	216.353	82.181
7	70.989	9.932	77.325	215.000	78.455
8	67.283	11.257	73.615	213.673	74.756
9	63.603	12.558	69.932	212.371	71.087
10	59.957	13.833	66.274	211.095	67.447
11	56.351	15.079	62.644	209.847	63.849
12	52.790	16.296	59.051	208.628	60.297
13	49.282	17.482	55.500	207.441	56.799
14	45.831	18.636	51.994	206.285	53.361
15	42.433	19.761	48.551	205.160	49.977
16	40.946	20.606	46.026	204.313	47.407
17	38.1789	21.449	43.350	203.469	44.826
18	35.000	22.439	41.027	79.859	42.433

TABLE 4.A.7

Case 7

Operating conditions:

Steam flow	137.073	[t/h]	Reject flow	6,297.180	[t/h]
Recycle flow	12,500.000	[t/h]	Makeup flow	6,142.800	[t/h]
Blowdown flow	5,056.901	[t/h]	Top brine temperature	95.000	[°C]
Product flow	1,082.199	[t/h]	Seawater flow	12,439.998	[t/h]
Performance ratio	7.775	[kg/540 kcal]			

Stages	TF_IN	D_OUT	TD_OUT	B_OUT	TB_OUT
1	85.245	1.191	90.541	207.109	91.583
2	81.926	2.363	87.216	205.930	88.257
3	78.618	3.520	83.912	204.772	84.954
4	75.329	4.658	80.615	203.631	81.666
5	72.055	5.779	77.342	202.509	78.397
6	68.801	6.882	74.084	201.404	75.148
7	65.569	7.967	70.848	200.318	71.920
8	62.360	9.033	67.631	199.250	68.716
9	59.178	10.080	64.439	198.203	65.540
10	56.024	11.107	61.271	197.174	62.392
11	52.908	12.113	58.128	196.167	59.281
12	49.831	13.096	55.019	195.182	56.211
13	46.799	14.056	51.947	194.221	53.186
14	43.817	14.991	48.915	193.284	50.213
15	40.878	15.905	45.937	192.369	47.283
16	39.608	16.599	43.708	191.675	45.042
17	37.432	17.271	41.462	191.001	42.855
18	35.000	18.037	39.555	84.282	40.878

TABLE 4.A.8

Case 8

Operating conditions:

Steam flow	148.918	[t/hr]	Reject flow	6,297.180	[t/h]
Recycle flow	12,500.000	[t/hr]	Makeup flow	6,142.800	[t/h]
Blowdown flow	4,966.366	[t/hr]	Top brine temperature	100.000	[°C]
Product flow	1,172.178	[t/hr]	Seawater flow	12,439.980	[t/h]
Performance ratio	7.831	[kg/540 kcal]			

Stages	TF_IN	D_OUT	TD_OUT	B_OUT	TB_OUT
1	89.484	1.296	95.219	206.996	96.290
2	85.887	2.573	91.615	205.713	92.683
3	82.298	3.832	88.033	204.452	89.100
4	78.726	5.070	84.457	203.210	85.530
5	75.170	6.289	80.903	201.990	81.979
6	71.634	7.488	77.365	200.789	78.449
7	68.120	8.666	73.849	199.609	74.940
8	64.632	9.823	70.354	198.451	71.457
9	61.171	10.959	66.886	197.315	68.005
10	57.743	12.072	63.444	196.200	64.582
11	54.354	13.160	60.029	195.110	61.199
12	51.009	14.224	56.651	194.045	57.862
13	47.713	15.261	53.313	193.006	54.575
14	44.473	16.270	50.021	191.996	51.345
15	41.282	17.254	46.786	191.011	48.166
16	39.967	17.996	44.393	190.267	45.747
17	37.626	18.714	41.973	189.549	43.393
18	35.000	19.536	39.936	82.773	41.282

TABLE 4.A.9

Case 9

Operating conditions:

Steam flow	162.173	[t/h]	Reject flow	6,297.180	[t/h]
Recycle flow	12,500.000	[t/h]	Makeup flow	6,142.800	[t/h]
Blowdown flow	4,876.161	[t/h]	Top brine temperature	105.000	[°C]
Product flow	1,261.744	[t/h]	Seawater flow	12,439.980	[t/h]
Performance ratio	7.873	[kg/540 kcal]			

Stages	TF_IN	D_OUT	TD_OUT	B_OUT	TB_OUT
1	93.724	1.402	99.896	206.884	100.99
2	89.852	2.781	96.018	205.495	97.116
3	85.985	4.142	92.160	204.133	93.255
4	82.134	5.481	88.306	202.790	89.405
5	78.296	6.798	84.474	201.471	85.574
6	74.479	8..093	80.656	200.174	81.762
7	70.684	9.364	76.861	198.901	77.972
8	66.915	10.612	73.088	197.651	74.210
9	63.175	11.836	69.344	196.427	70.481
10	59.469	13.035	65.626	195.226	66.781
11	55.807	14.206	61.938	194.053	63.125
12	52.192	15.349	58.290	192.909	59.519
13	48.631	16.462	54..686	191.794	55.968
14	45.131	17.545	51.131	190.710	52.481
15	41.687	18.599	47.639	189.655	49.050
16	40.325	19.389	45.079	188.864	46.454
17	37.819	20.151	42.484	188.101	43.932
18	35.000	21.029	40.317	81.269	41.687

TABLE 4.A.10

Case 10

Operating conditions:

Steam flow	123.173	[t/h]	Reject flow	6,297.180	[t/h]
Recycle flow	11,500.000	[t/h]	Makeup flow	6,142.800	[t/h]
Blowdown flow	5,131.935	[t/h]	Top brine temperature	95.000	[°C]
Product flow	1,007.189	[t/h]	Seawater flow	12,439.998	[t/h]
Performance ratio	8.041	[kg/540 kcal]			

Stages	TF_IN	D_OUT	TD_OUT	B_OUT	TB_OUT
1	85.331	1.116	90.448	190.516	91.515
2	81.954	2.215	87.103	189.411	88.128
3	78.588	3.299	83.740	188.326	84.767
4	75.241	4.364	80.386	187.258	81.421
5	71.911	5.414	77.056	186.207	78.096
6	68.603	6.445	73.742	185.174	74.791
7	65.316	7.460	70.451	184.158	71.508
8	62.055	8.456	67.181	183.160	68.252
9	58.821	9.435	63.938	182.180	65.024
10	55.618	10.394	60.719	181.219	61.826
11	52.453	11.333	57.528	180.279	58.666
12	49.330	12.251	54.373	179.360	55.549
13	46.254	13.146	51.256	178.464	52.480
14	43.229	14.018	48.182	177.590	49.464
15	40.251	14.870	45.163	176.738	46.495
16	39.077	15.514	42.897	176.092	44.224
17	37.126	16.117	40.726	175.489	42.093
18	35.000	16.786	38.957	85.532	40.251

TABLE 4.A.11

Case 11

Operating conditions:

Steam flow	133.780	[t/h]	Reject flow	6,297.180	[t/h]
Recycle flow	11,500.000	[t/h]	Makeup flow	6,142.800	[t/h]
Blowdown flow	5,047.659	[t/h]	Top brine temperature	100.000	[°C]
Product flow	1,090.915	[t/h]	Seawater flow	12,439.980	[t/h]
Performance ratio	8.1	[kg/540 kcal]			

Stages	TF_IN	D_OUT	TD_OUT	B_OUT	TB_OUT
1	89.578	1.216	95.161	190.410	96.215
2	85.918	2.412	91.492	189.207	92.543
3	82.266	3.591	87.847	188.026	88.897
4	78.633	4.750	84.208	186.863	85.264
5	75.016	5.891	80.593	185.721	81.652
6	71.421	7.013	76.995	184.597	78.061
7	67.849	8.115	73.420	183.494	74.493
8	64.304	9.196	69.868	182.411	70.954
9	60.787	10.257	66.344	181.349	67.446
10	57.305	11.297	62.847	180.308	63.968
11	53.864	12.313	59.380	179.290	60.533
12	50.469	13.305	55.952	178.297	57.146
13	47.126	14.272	52.566	177.328	53.811
14	43.840	15.213	49.228	176.386	50.535
15	40.607	16.129	45.950	175.469	47.313
16	39.397	16.820	43.518	174.778	44.865
17	37.298	17.462	41.179	174.135	42.572
18	35.000	18.182	39.293	84.128	40.607

TABLE 4.A.12

Case 12

Operating conditions:

Steam flow	145.656	[t/h]	Reject flow	6,297.180	[t/h]
Recycle flow	11,500.000	[t/h]	Makeup flow	6,142.800	[t/h]
Blowdown flow	4,963.671	[t/h]	Top brine temperature	105.000	[°C]
Product flow	1,174.270	[t/h]	Seawater flow	12,439.980	[t/h]
Performance ratio	8.144	[kg/540 kcal]			

Stages	TF_IN	D_OUT	TD_OUT	B_OUT	TB_OUT
1	93.827	1.314	99.834	190.304	100.918
2	89.887	2.607	95.886	189.003	96.964
3	85.952	3.882	91.960	187.726	93.36
4	82.034	5.135	88.039	186.469	89.118
5	78.132	6.368	84.140	185.234	85.221
6	74.250	7.579	80.257	184.021	81.344
7	70.392	8.768	76.399	182.829	77.491
8	66.563	9.935	72.564	181.662	73.668
9	62.763	11.078	68.760	180.518	69.878
10	58.999	12.198	64.983	179.396	66.121
11	55.281	13.291	61.239	178.301	62.409
12	51.612	14.435	57.538	177.233	58.748
13	48.001	15.395	53.882	176.194	55.146
14	44.452	16.404	50.278	175.184	51.610
15	40.963	17.386	46.740	174.202	48.134
16	39.715	18.120	44.141	173.467	45.507
17	37.469	18.802	41.634	172.784	43.051
18	35.000	19.571	39.628	82.724	40.963

References

Cameron, M.C. (1996), Data reconciliation—Progress and challenges, *J. Process Control*, **6**(2/3), 89–98.

Liebman, M.J., T.F. Edgar, and L.S. Lasdon (1992), Efficient data reconciliation and estimation for dynamic process using non linear programming techniques, *Comput. Chem. Eng.*, **16**, 963–986.

5

Analysis of the Dynamic Model for Control

5.1 Introduction

Control engineering is a composite technical activity requiring several functions to be performed in order to render it effective and excellent. These functions include modeling, simulation, instrumentation, system design, data handling, and so on. Each of these functions has its own role to play in efficiently meeting the objectives of control, which, too, has to conform to the objectives of the overall process performance. For example, the need to provide the process information for the control system is met by the measurement and data acquisition system. This information or data comprising all signals relevant to the context of control has to be free from errors that occur in reality. The removal or minimization of the influence of such errors is to be done by suitable processing of the measured raw data.

The subject of data purification is vast. The field of estimation, which embodies techniques of smoothing, filtering, and prediction, is closely allied to the area of control. Uncertain and noisy situations in practice require the application of stochastic control in which estimation is an essential ingredient. At this stage of the present work, a detailed consideration of the field of estimation is not intended. However, for the purpose of purifying the measured (raw) data in real processes, a technique that is in the spirit of the methods of estimation should be discussed as a prerequisite to our further investigations. This procedure is known as *data reconciliation* in the process engineering field to minimize the influence of measurement errors mainly in the heat and mass variables (Cameron, 1996). Data reconciliation is a vital issue since erroneous measurements due to malfunction or dislocation of sensors can be countered and minimized. Data obtained from a process always include measurement errors. Measurement errors occur due to the device itself or due to the installation of a device. Therefore, data reconciliation ensures that the measured data satisfy the condition such as mass and energy balance in the process. This is done by minimizing an objective function of the error between the measured and reconciled values subject to physical balance conditions as illustrated in Chapter 4.

The interrelationships among the various procedures in the wide setup of plant measurement, information processing, control, and so forth are evident from Figure 5.1. The process variables are sensed and passed on to the data-processing stage after proper conditioning through the data acquisition unit. The data-processing stage embodies all procedures such as validation, filtering, reconciliation, coaptation, and so on, to purify the measured signals, which are usually subjected to random errors and disturbances. The purified process variables may be used for various purposes such as monitoring, identification, feedback, and so on. In the identification procedure, the input and output data are fitted to a model whose structure is chosen on the basis of available a priori information and the estimated model parameters will be made available for procedures such as controller design, setpoint optimization, and so on. The controller design stage should also be provided with control performance specifications. The controller parameters are then set according to the results of the controller design stage. The controller output or the manipulated variable may not always be rich enough in its information content for successful identification. Therefore, a test signal may be injected if necessary as shown in Figure 5.1.

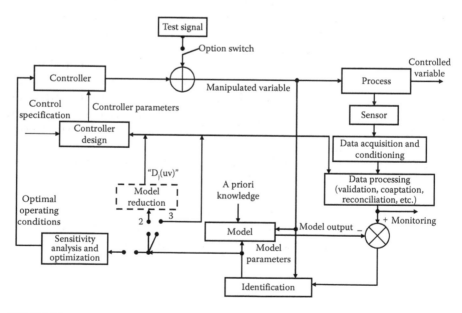

FIGURE 5.1
The wide setup of process control with various functions.

5.2 Selection of Sensors, Valves, and Controlled and Manipulated Variables

One of the most important aspects of process control is to identify the variables that are to be measured for the purpose of control studies. It is necessary to know the input and output characteristics of the measurement elements in order to fully design a control loop. These issues relate primarily to the selection of control structures, placement of sensors, and pairing those sensors with available manipulated variables within that structure. Proper choices at this level typically have a far greater impact on the overall performance of the control system than does the choice of control algorithms.

Important considerations in the selection of the sensors, valves, and controlled and manipulated variables are as follows (Luyben, 1992):

Sensors: Issues such as analyzer type and focus are important in multicomponent distillation process, besides choosing proper location of control sensors, sensitivity, consistency, reliability, etc. The location of the sensors is based on a study of loop sensitivity. Numerically efficient singular value decomposition (SVD)-based methods are used in finding the most sensitive location for sensors (Luyben, 1992).

Valves: There are many valve issues to be considered in the design of a control strategy, that is, the valve range, linearity, the size of a control values, dynamics, etc.

Controlled variables: These are the key variables that govern the operation of a process. These are usually regulated at chosen levels to maintain the process at the desired point of operation.

Manipulated variables: These are the variables that are manipulated to regulate the controlled variables. The proper pairing scheme is the result of the so-called relative gain array (RGA) analysis (Seborg, 2010) whose objective is to arrive at pairing with large leverage for control and negligible interaction. In all these situations, SVD-based methods are used.

The open-loop process characteristics are the basis for control system design. In Chapter 3, we obtained such a model for the MSF plant. This is usually described by a set of nonlinear ordinary differential equations (ODEs) based on lumped approximations of distributed parameter phenomena and nonlinear correlations among the process parameters and variables. These equations may be linearized about any chosen operating condition. The result is a linearized model of the process, which is useful in the design of a controller. A linear model in state space suitable for control design can be derived from it. This work considers an 18-stage MSF process modeled and simulated using the SPEEDUP package. The process is simulated at different conditions of operation, and the results of linearization of the dynamics are examined. The conditions of operation correspond to different loads rates of production of distillate. The study provides an insight into the complexity and nonlinearity of the dynamic model, forming a basis for advanced control proposals.

5.3 Model Linearization

A process is generally described by a set of nonlinear differential and algebraic equations. The differential equations are obtained in ordinary form by lumped approximation of the distributed process, which actually is better described by partial differential equations (PDEs). Lumped approximations reduce the PDEs into ODEs. However, these equations are usually nonlinear, since they describe nonlinear phenomena. There are no techniques generally applicable to nonlinear systems except numerical solutions to the cases in hand. The design of control systems for nonlinear processes is usually made to meet the specifications in a regulatory mode rather than in the tracking mode. That is, the process is normally to work at fixed operating conditions. It is, therefore, reasonable to consider the close neighborhood of the operating point to linearize the plant model and then use the well-established linear control methodology to design the related control system. Therefore, the first step in our further studies is model linearization about a chosen operating point, which is as follows:

Consider the set of nonlinear differential (ordinary) and algebraic equations generally used that describe a process:

$$\dot{\mathbf{x}}(t) = \mathbf{f}\big[\mathbf{x}(t), \quad \mathbf{u}(t)\big] \tag{5.1}$$

$$\mathbf{y}(t) = \mathbf{g}\big[\mathbf{x}(t), \quad \mathbf{u}(t)\big] \tag{5.2}$$

where

$$\mathbf{x}(t) = \big[x_1(t) \quad x_2(t) \quad \cdots \quad x_n(t)\big]^{\mathrm{T}} \tag{5.3}$$

$$\mathbf{y}(t) = \big[y_1(t) \quad y_2(t) \quad \cdots \quad y_m(t)\big]^{\mathrm{T}} \tag{5.4}$$

$$\mathbf{u}(t) = \big[u_1(t) \quad u_2(t) \quad \cdots \quad u_r(t)\big]^{\mathrm{T}} \tag{5.5}$$

and $\mathbf{f}$ and $\mathbf{g}$ are n and m vector-valued functions of the state $\mathbf{x}(t)$ and the inputs $\mathbf{u}(t)$. Equations 5.1 and 5.2 are known as the state and output equations, respectively.

Consider a steady-state operating condition $\bar{x}$, $\bar{u}$ and let the process be perturbed by small signals $x^*(t)$ and $u^*(t)$ such that

$$x(t) = \bar{x} + x^*(t)$$
$$u(t) = \bar{u} + u^*(t)$$

(5.6)

Equation 5.1 is linearized to appear in the form

$$x^*(t) = Ax^*(t) + Bu^*(t)$$

(5.7)

where

$$A = \begin{bmatrix} \dfrac{\partial f_1}{\partial x_1} & \cdots & \dfrac{\partial f_1}{\partial x_n} \\ \cdot & \cdot & \cdot \\ \cdot & \cdot & \cdot \\ \dfrac{\partial f_n}{\partial x_1} & \cdots & \dfrac{\partial f_n}{\partial x_n} \end{bmatrix}_{\substack{x=\bar{x} \\ u=\bar{u}}}$$

(5.8)

and

$$B = \begin{bmatrix} \dfrac{\partial f_1}{\partial u_1} & \cdots & \dfrac{\partial f_1}{\partial u_r} \\ \cdot & \cdot & \cdot \\ \cdot & \cdot & \cdot \\ \dfrac{\partial f_n}{\partial u_1} & \cdots & \dfrac{\partial f_n}{\partial u_r} \end{bmatrix}_{\substack{x=\bar{x} \\ u=\bar{u}}}$$

(5.9)

Similarly, Equation 5.2 is linearized into the form

$$y^*(t) = Cx^*(t) + Du^*(t)$$

(5.10)

where

$$C = \begin{bmatrix} \dfrac{\partial g_1}{\partial x_1} & \cdots & \dfrac{\partial g_1}{\partial x_n} \\ \cdot & \cdot & \cdot \\ \cdot & \cdot & \cdot \\ \dfrac{\partial g_m}{\partial x_1} & \cdots & \dfrac{\partial g_m}{\partial x_n} \end{bmatrix}_{\substack{x=\bar{x} \\ u=\bar{u}}}$$

(5.11)

and

$$D = \begin{bmatrix} \dfrac{\partial g_1}{\partial u_1} & \cdots & \dfrac{\partial g_1}{\partial u_r} \\[2ex] \cdot & \cdot & \cdot \\ \cdot & \cdot & \cdot \\[1ex] \dfrac{\partial g_m}{\partial u_1} & \cdots & \dfrac{\partial g_m}{\partial u_r} \end{bmatrix}_{\substack{x=\bar{x}\\u=\bar{u}}} \qquad (5.12)$$

For notational simplicity, Equations 5.7 and 5.10 are written without the asterisk as

$$\dot{x} = Ax + Bu \qquad (5.13)$$

$$y = Cx + Du \qquad (5.14)$$

These are the linearized state-space equations about $(\bar{x}, \bar{u})$. Control system design is based on these equations. The linear model and the related controller are valid in the close neighborhood of the operating point $(\bar{x}, \bar{u})$.

5.4 Linearized Model of an MSF Plant

The model of an 18-stage MSF plant is set up in SPEEDUP flowsheet simulator. Twelve operating conditions (cases) as listed in Tables 4.A.1 through 4.A.12 have been chosen, and the model is linearized at each of these conditions using a dynamic run and invoking a control design interface (CDI) of SPEEDUP with the following six inputs and six outputs (Woldai et al., 1996):

Manipulated variables (inputs):

u_1: Culvert flow controller valve position
u_2: Makeup flow controller valve position
u_3: Recycle flow control valve position
u_4: Seawater recirculation flow control valve position
u_5: Reject flow control valve position
u_6: Steam flow control to the brine heater

Controlled variables (outputs):

y_1: Top brine temperature of the brine heater (TBT)

y_2: Culvert flow

y_3: Seawater feed flow entering the last flash (F18.flow)

y_4: Brine recycle flow (F18.Recycle flow)

y_5: Temperature of the seawater entering the last flash

y_6: Makeup flow entering the last flash

The CDI of SPEEDUP generates the matrices **A**, **B**, **C**, and **D** of the state space and computes the steady-state gain matrix **G**(0).

The present linearized model has 155 state variables in which 7 state variables are in each flash stage (7*18) and the rest are in the brine heater and controllers.

From the linear state-space description Equations 5.13 and 5.14 of a multi-input–multi-output (MIMO) process, the transfer function matrix between the inputs (manipulated variables) and the outputs (controlled variables) is given as

$$G(s) = C(sI - A)^{-1}B + D \qquad (5.15)$$

The elements of **D** are usually zeros. But if the output y_j is directly proportional to u_i, without involving any dynamics in between,

$$y_j = k_{ij}u_i$$

then k_{ij} will be nonzero in **D**. This will remain as it is to appear in **G**(s) and in **G**(0).

In this work, it will be seen that there are apparently direct, nondynamic, couplings between some inputs and outputs within the system. These couplings are shown by the nonzero output responses with no dynamic component within the time histories shown. It should be noted that, in reality, there is a dynamic component to this coupling. However, the speed of these dynamics is so fast as to appear instantaneous at the sample rate chosen for this modeling work.

The steady-state gain matrix or the DC gain matrix for nonintegrating processes can be computed at $s = 0$ in the aforementioned equation as

$$G(0) = -CA^{-1}B \qquad (5.16)$$

It should be noted that in the case of integrating processes, that is, those having poles at the origin of the s-plane (usually due to level control systems), A^{-1} does not exist. We will not consider such situations here.

The following are the DC gain matrices at the chosen operating conditions ($G_{TBT,REC}(0)$, called Cases 1–12, defined as

	Recycle Flow (t/h)			
TBT (°C)	**14,420**	**13,500**	**12,500**	**11,500**
95	Case 1	Case 4	Case 7	Case 10
100	Case 2	Case 5	Case 8	Case 11
105	Case 3	Case 6	Case 9	Case 12

where TBT, temperature in °C; REC, recycle flow in t/h:

$$
\text{Case 1}: G_{95,14.42}(0) =
\begin{bmatrix}
0.0000 & -2.1884 & -82.0318 & 0.1060 & -2.8008 & 54.0022 \\
61.700 & 176.2433 & 0.0000 & 0.0000 & 183.3005 & 0.0000 \\
0.0000 & 176.2433 & 0.0000 & 96.7000 & 183.3005 & 0.0000 \\
0.0000 & 0.0000 & 541.0021 & 0.0000 & 0.0000 & 0.0000 \\
0.0000 & 0.0000 & 0.0000 & 3.4966 & 0.0000 & 0.0000 \\
0.0000 & 176.2433 & 0.0000 & 0.0000 & 0.0000 & 0.0000
\end{bmatrix}
$$

$$
\text{Case 2}: G_{100,14.42}(0) =
\begin{bmatrix}
0.0000 & -1.8275 & -68.7636 & 0.0862 & -2.3144 & 39.8536 \\
61.700 & 176.2433 & 0.0000 & 0.0000 & 183.3005 & 0.0000 \\
0.0000 & 176.2433 & 0.0000 & 96.7000 & 183.3005 & 0.0000 \\
0.0000 & 0.0000 & 541.0021 & 0.0000 & 0.0000 & 0.0000 \\
0.0000 & 0.0000 & 0.0000 & 3.7680 & 0.0000 & 0.0000 \\
0.0000 & 176.2433 & 0.0000 & 0.0000 & 0.0000 & 0.0000
\end{bmatrix}
$$

$$
\text{Case 3}: G_{105,14.42}(0) =
\begin{bmatrix}
0.0000 & -1.3565 & -51.5614 & 0.0622 & -1.6959 & 22.0534 \\
61.700 & 176.2433 & 0.0000 & 0.0000 & 183.3005 & 0.0000 \\
0.0000 & 176.2433 & 0.0000 & 96.700 & 183.3005 & 0.0000 \\
0.0000 & 0.0000 & 541.0021 & 0.0000 & 0.0000 & 0.0000 \\
0.0000 & 0.0000 & 0.0000 & 4.0395 & 0.0000 & 0.0000 \\
0.0000 & 176.2433 & 0.0000 & 0.0000 & 0.0000 & 0.0000
\end{bmatrix}
$$

$$
\text{Case 4}: G_{95,13.50}(0) =
\begin{bmatrix}
0.0000 & -1.9764 & -83.5318 & 0.1075 & -2.6048 & 56.2731 \\
61.7000 & 0.0000 & 0.0000 & 0.0000 & 0.0000 & 0.0000 \\
0.0000 & 176.2433 & 0.0000 & 96.7000 & 183.3010 & 0.0000 \\
0.0000 & 0.0000 & 506.4862 & 0.0000 & 0.0000 & 0.0000 \\
0.0000 & 0.0000 & 0.0000 & 3.2231 & 0.0000 & 0.0000 \\
0.0000 & 176.2433 & 0.0000 & 0.0000 & 0.0000 & 0.0000
\end{bmatrix}
$$

$$\text{Case 5}: G_{100,13.50}(0) = \begin{bmatrix} 0.0000 & -1.6599 & -70.4041 & 0.0881 & -2.1678 & 42.6115 \\ 61.7000 & 0.0000 & 0.0000 & 0.0000 & 0.0000 & 0.0000 \\ 0.0000 & 176.2433 & 0.0000 & 96.7000 & 183.3010 & 0.0000 \\ 0.0000 & 0.0000 & 506.4862 & 0.0000 & 0.0000 & 0.0000 \\ 0.0000 & 0.0000 & 0.0000 & 3.4732 & 0.0000 & 0.0000 \\ 0.0000 & 176.2433 & 0.0000 & 0.0000 & 0.0000 & 0.0000 \end{bmatrix}$$

$$\text{Case 6}: G_{105,13.5}(0) = \begin{bmatrix} 0.0000 & -1.2450 & -53.3271 & 0.0643 & -1.6077 & 25.3382 \\ 61.700 & 176.2433 & 0.0000 & 0.0000 & 183.3005 & 0.0000 \\ 0.0000 & 176.2433 & 0.0000 & 96.700 & 183.3005 & 0.0000 \\ 0.0000 & 0.0000 & 506.4860 & 0.0000 & 0.0000 & 0.0000 \\ 0.0000 & 0.0000 & 0.0000 & 3.7235 & 0.0000 & 0.0000 \\ 0.0000 & 176.2433 & 0.0000 & 0.0000 & 0.0000 & 0.0000 \end{bmatrix}$$

$$\text{Case 7}: G_{95,12.50}(0) = \begin{bmatrix} 0.0000 & -1.7434 & -84.9278 & 0.1085 & -2.3826 & 58.6086 \\ 61.7000 & 0.0000 & 0.0000 & 0.0000 & 0.0000 & 0.0000 \\ 0.0000 & 176.2433 & 0.0000 & 96.7000 & 183.3010 & 0.0000 \\ 0.0000 & 0.0000 & 468.9679 & 0.0000 & 0.0000 & 0.0000 \\ 0.0000 & 0.0000 & 0.0000 & 2.9287 & 0.0000 & 0.0000 \\ 0.0000 & 176.2433 & 0.0000 & 0.0000 & 0.0000 & 0.0000 \end{bmatrix}$$

$$\text{Case 8}: G_{100,12.50}(0) = \begin{bmatrix} 0.0000 & -1.4713 & -71.9948 & 0.0895 & -1.9964 & 45.4694 \\ 61.7000 & 0.0000 & 0.0000 & 0.0000 & 0.0000 & 0.0000 \\ 0.0000 & 176.2433 & 0.0000 & 96.7000 & 183.3010 & 0.0000 \\ 0.0000 & 0.0000 & 468.9679 & 0.0000 & 0.0000 & 0.0000 \\ 0.0000 & 0.0000 & 0.0000 & 3.1559 & 0.0000 & 0.0000 \\ 0.0000 & 176.2433 & 0.0000 & 0.0000 & 0.0000 & 0.0000 \end{bmatrix}$$

$$\text{Case 9}: G_{105,12.5}(0) = \begin{bmatrix} 0.0000 & -1.1162 & -55.1290 & 0.0663 & -1.5003 & 28.7852 \\ 61.700 & 176.2433 & 0.0000 & 0.0000 & 183.3005 & 0.0000 \\ 0.0000 & 176.2433 & 0.0000 & 96.700 & 183.3005 & 0.0000 \\ 0.0000 & 0.0000 & 468.9684 & 0.0000 & 0.0000 & 0.0000 \\ 0.0000 & 0.0000 & 0.0000 & 0.0000 & 0.0000 & 0.0000 \\ 0.0000 & 176.2433 & 0.0000 & 0.0000 & 0.0000 & 0.0000 \end{bmatrix}$$

$$\text{Case 10}: G_{95,11.50}(0) = \begin{bmatrix} 0.0000 & -1.5134 & -86.1961 & 0.1089 & -2.1551 & 60.9066 \\ 61.7000 & 0.0000 & 0.0000 & 0.0000 & 0.0000 & 0.0000 \\ 0.0000 & 176.2433 & 0.0000 & 96.7000 & 183.3010 & 0.0000 \\ 0.0000 & 0.0000 & 431.4497 & 0.0000 & 0.0000 & 0.0000 \\ 0.0000 & 0.0000 & 0.0000 & 2.6376 & 0.0000 & 0.0000 \\ 0.0000 & 176.2433 & 0.0000 & 0.0000 & 0.0000 & 0.0000 \end{bmatrix}$$

$$\text{Case 11}: G_{100,11.50}(0) = \begin{bmatrix} 0.0000 & -1.2815 & -73.4424 & 0.0904 & -1.8167 & 48.2383 \\ 61.7000 & 0.0000 & 0.0000 & 0.0000 & 0.0000 & 0.0000 \\ 0.0000 & 176.2433 & 0.0000 & 96.7000 & 183.3010 & 0.0000 \\ 0.0000 & 0.0000 & 431.4497 & 0.0000 & 0.0000 & 0.0000 \\ 0.0000 & 0.0000 & 0.0000 & 2.8424 & 0.0000 & 0.0000 \\ 0.0000 & 176.2433 & 0.0000 & 0.0000 & 0.0000 & 0.0000 \end{bmatrix}$$

$$\text{Case 12}: G_{105,11.5}(0) = \begin{bmatrix} 0.0000 & -0.9817 & -56.7690 & 0.0677 & -1.3815 & 32.0900 \\ 61.700 & 176.2433 & 0.0000 & 0.0000 & 183.3005 & 0.0000 \\ 0.0000 & 176.2433 & 0.0000 & 96.700 & 183.3005 & 0.0000 \\ 0.0000 & 0.0000 & 431.4492 & 0.0000 & 0.0000 & 0.0000 \\ 0.0000 & 0.0000 & 0.0000 & 3.0473 & 0.0000 & 0.0000 \\ 0.0000 & 176.2433 & 0.0000 & 0.0000 & 0.0000 & 0.0000 \end{bmatrix}$$

5.5 Control Structure

Bristol (1966) developed a systematic approach for the analysis of multivariable process control problems. His approach requires only steady-state information and provides two important items of information:

1. A measure of process interaction
2. A recommendation concerning the most effective pairing of controlled and manipulated variables

Bristol's approach is based on the concept of a relative gain. RGA Λ is a square matrix (for equal number of manipulated and controller variables) whose columns refer to manipulated variables and the rows to controlled variables. That is,

$$\Lambda = \begin{bmatrix} \lambda_{11} & \lambda_{12} & \cdots & \lambda_{1n} \\ \lambda_{21} & \lambda_{22} & \cdots & \lambda_{2n} \\ \cdot & \cdot & \cdot & \cdot \\ \cdot & \cdot & \cdot & \cdot \\ \cdot & \cdot & \cdot & \cdot \\ \lambda_{n1} & \lambda_{n2} & \cdots & \lambda_{nn} \end{bmatrix} \tag{5.17}$$

The RGA Λ has two important properties:

1. It is normalized, since the sum of the elements in each row or column is one.

$$\sum_{i=1}^{n} \lambda_{ij} = 1, \quad \text{for all} \quad j = 1, 2, \ldots, n$$

2. The relative gains are dimensionless and thus not affected by choice of units or scaling of variables: λ_{ij} is dimensionless.

Relative gain values close to or equal to one indicate that closing or opening other loops does not have an effect on loop $j - i$, which in this case is a preferable pairing. Values close to zero indicate that input j only has a negligible effect on the output i; obviously, such pairings cannot achieve good control. For $0 < \lambda_{ij} < 1$, the open-loop gain is lower than the closed-loop gain, implying that interaction between the loops exists (maximal interaction at $\lambda_{ij} = 0.5$); the same holds for values greater than one. Pairing with negative relative gain values should be avoided in any case, since they imply that the open-loop effect of a manipulated variable on an output is reversed if other loops are operated; in case of failure of one of the loops, a control scheme implying negative relative gains could become unstable. If the input and output vectors are arranged in the order of their pairing, the target should be RGA as close to the identity matrix as possible. If no structure with acceptable decoupling can be identified, either a multivariable controller or a decoupling design should be considered.

Consider the present system with six inputs and six outputs. The steady-state gain matrix $G_{95,14,92}(0)$ for a 6×6 system is shown in the following:

$$\begin{bmatrix} y_1 \\ y_2 \\ y_3 \\ y_4 \\ y_5 \\ y_6 \end{bmatrix} = \begin{bmatrix} 0.0000 & -2.1884 & -82.0318 & 0.1060 & -2.8008 & 54.0022 \\ 61.700 & 176.2433 & 0.0000 & 0.0000 & 183.3005 & 0.0000 \\ 0.0000 & 176.2433 & 0.0000 & 96.7000 & 183.3005 & 0.0000 \\ 0.0000 & 0.0000 & 541.0021 & 0.0000 & 0.0000 & 0.0000 \\ 0.0000 & 0.0000 & 0.0000 & 3.4966 & 0.0000 & 0.0000 \\ 0.0000 & 176.2433 & 0.0000 & 0.0000 & 0.0000 & 0.0000 \end{bmatrix} \begin{bmatrix} u_1 \\ u_2 \\ u_3 \\ u_4 \\ u_5 \\ u_6 \end{bmatrix}$$

From the DC gain matrix, Λ can be computed by using the relation:

$$\lambda_{ij} = \frac{(\partial y_i / \partial u_j)_{u_1 \neq j}}{(\partial y_i / \partial u_j)_{y_1 \neq i}} = g_{ij}[G^{-1}]'_{ij} \tag{5.18}$$

Manipulated and controlled variables are so paired that the relative gains λ_{ij} are as close to unity as possible. That is, (controlled variable)$_i$ with (manipulated variable)$_j$ should be paired if λ_{ij} is close to 1. In all the cases here

$$\Lambda = \begin{bmatrix} 0 & 0 & 0 & 0 & 0 & 1 \\ 1 & 0 & 0 & 0 & 0 & 0 \\ 0 & 0 & 0 & 0 & 1 & 0 \\ 0 & 0 & 1 & 0 & 0 & 0 \\ 0 & 0 & 0 & 1 & 0 & 0 \\ 0 & 1 & 0 & 0 & 0 & 0 \end{bmatrix}$$

clearly suggesting pairing as follows: $(u_1 \rightarrow y_2)$, $(u_2 \rightarrow y_6)$, $(u_3 \rightarrow y_4)$, $(u_4 \rightarrow y_5)$, $(u_5 \rightarrow y_3)$, and $(u_6 \rightarrow y_1)$, which means that they have insignificant interaction from the other loops. Thus, they can be tuned independently using the conventional single-loop tuning methods.

RGA analysis should also take into account the Niederlinski test for stability and is better done with $G(j\omega)$ over a range of ω around the Nyquist crossover point. Thus, the control structure emerges out of the RGA analysis. Since the RGA analysis in all the cases here indicates a very clear pairing strategy, we need no further tests.

The transfer function matrices have high-order elements that require reduction. This task is performed in the next chapter. For now, we can see the dynamic behavior in Figure 6.A.1a through l, which provides a picture of the step response matrices in the 12 cases of the present investigation. Time is reckoned in minutes in these figures. Notice that some elements of the transfer function matrix have a negligible magnitude, while some others are without dynamics.

5.6 Remarks

The dynamic model of the MSF plant obtained here exhibits nonlinear behavior to some extent. This is evident from the variation of the parameters of the linearized model, namely, the transfer function matrix. The transfer function G_{16} is of much importance in the control of the MSF plant. In the first place, since the TBT is critical for the plant performance, its control is important. Second, the effectiveness of TBT control renders the plant operation effective; good TBT control makes the plant operation steady with minimal risk of flooding or blowthrough under changing settings.

The steady-state or DC gain of G_{16} decreases with the TBT increase at fixed recycle flow. This is evident from the nonlinearity of the LP steam valve characteristic. The inherent nonlinearity of the brine heater behavior, which is the result of nonlinear correlations among the thermal parameters, may be another reason. This is difficult to visualize in view of the complexity of the model description. The DC gain of G_{16} also with the increasing recycle flowrate. This is a well-known phenomenon in heat exchangers in which the DC gain varies inversely as the rate of flow of the heat receiving stream.

The elements $G_{ij}(s)$ of $\mathbf{G}(s)$ are of an order 155, which is too high to render them tractable for control design considerations. The order of these elements should be reduced in an appropriate manner so that further studies for control system design can be taken up. Thus, the task immediately ahead is that of model reduction, which will be discussed in Chapter 6.

References

Bristol, E.M. (1966), On a new measure of interaction for multivariable process control, *IEEE Trans. Automat. Control*, **AC-11**, 133.

Cameron, M.C. (1996), Data reconciliation—Progress and challenges, *J. Process Control*, **6**(2/3), 89–98.

Luyben, W.L. (1992), *Practical Distillation Control*, Van Nostrand, New York.

Seborg, D.E., D.A. Mellichamp, T.F. Edgar, and F.J. Doyle, III (2010), *Process Dynamics and Control, 3rd Edition*, John Wiley and Sons, Hoboken, NJ.

Woldai, A., D.M.K. Al-Gobaisi, R.W. Dunn, A. Kurdali, and G.P. Rao (1996), An adaptive scheme with an optimally tuned PID controller for a large MSF desalination plant, *Control Eng. Pract.*, **4**(5), 721–734.

6

Optimal Reduction of the MSF Plant Model to the First or Second Order with Delay Forms

The model of the multi-stage flash (MSF) plant obtained in the previous chapter is too large and complex to be useful for control design. Model reduction techniques are often required in control system analysis for obtaining low-order approximations of high-order systems. Using time-domain concepts, many reduction methods have been proposed and recently, much work has utilized the idea of balanced realizations introduced by Moore (1981) and balanced truncation by Safanov and Chiang (1989), and so on. Considering frequency-domain techniques, a wide variety of reduction methods have been suggested by Jamshidi (1983) and others.

The linearized model of an MSF plant is large and complex in size, requiring reductions for controller design and practical implementation. This chapter includes a brief review of model reduction techniques. The need for modeling controlled processes in simple forms such as first order or second order with delay forms largely arises out of the widely prevailing PID control practice, with possibilities for advanced features such as optimization and adaptability. The case of our 18-stage MSF desalination plant model is subjected to the studies in this direction.

6.1 Model Reduction Methods

The main objective of model reduction is to obtain a reduced-order approximation of a complex high-order system that retains and reflects the important characteristics of the original system as closely as possible.

Some of the reasons for using reduced-order models of high-order systems include the following:

- To have a better understanding of the system. A system of uncomfortably high order poses difficulties in its analysis, simulation, synthesis, or identification. An obvious method for dealing with such systems is to approximate then to a low-order system for which characteristics such as time constant, damping ratio, natural frequency, and their interrelationships are well known.

- To reduce computational burden.
- To simplify control systems design.
- To make feasible designs. Reduced-order models may be used effectively in situations like
  - Model reference adaptive control systems
  - Online interactive system modeling
  - Decentralized controllers
  - Hierarchical control systems, etc.

Model-order reduction has been attempted both in the time and in the frequency domains. The time-domain methods are commonly associated with state-space models, while the frequency-domain methods are associated with the transfer function or transfer matrix representations. Some important frequency-domain methods are the Padé approximation, continued fraction expansion, moment matching, Routh approximation, and so on. The time-domain methods include balanced realization, Hankel-based methods, singular perturbation, optimal methods, the aggregation technique, and so on.

A detailed review of the field of model reduction is beyond the scope of this chapter. The area of model reduction is dominated by methods generally aimed at obtaining finite dimensional approximations either in state-space or in the complex frequency domain. Inclusion of a delay term in the reduced model is generally rare. Some of the methods are discussed in this chapter.

6.1.1 Time-Domain Methods

The time-domain technique may be classified into three groups: (1) modal methods, (2) optimal methods, and (3) simple thumb rules and graphical methods.

The modal methods by Marshall (1966) retain the dominant eigenvalues of the original system and then obtain the remaining unknown parameters of the low-order model such that its response to certain inputs approximates closely to that of the high-order system. The optimal methods by Wilson and Mishra (1979) are based on obtaining a low-order model of a given order so that its impulse (or step) response will match that of the original system in an optimum manner, with no restriction on the location of the eigenvalues. Such methods normally aim to minimize a selected performance criterion and are, therefore, mathematically more rigorous. The performance criterion chosen is some function of the error between the time responses of the original systems and its reduced version. There is the thumb rule: for example, in order to obtain a first-order plus delay approximation, based on two instants of time from the step response, namely,

1. $t_{28.3}$: Time to attain 28.3% of the steady-state value
2. $t_{63.2}$: Time to attain 63.2% of the steady-state value

Then, the time constant of the approximate first-order lag are taken as $1.5(t_{63.2} - t_{28.3})$ and the delay is taken as $(t_{63.2} - \text{time constant})$.

There are several graphical methods that are described in detail by Unbehauen and Rao (1987). They use tangents at the points of inflection along with the results of other graphical procedures.

6.1.2 State-Space Methods

The first step in model reduction in state space is to remove the redundant states. This is aimed at obtaining the minimal realization. Recall that the transfer function matrix (TFM) of the system is given by

$$G(s) = C(sI - A)^{-1}B + D$$

$G(s)$ can be computed by Le verrier type of algorithm. The result is TFM, all the elements of which have a common denominator called the *characteristic polynomial*. If the elements of the TFM are all made coprime by pole zero cancellation, we get the minimal realization. The cancellation procedure is applied under certain tolerance limit.

In the state space, the minimal realization is achieved by first obtaining the controllability and observability staircase forms in which the uncontrollable and unobservable states are isolated and removed.

Let n be the size of A. If the controllability matrix of the pair (A, B) has rank $r \leq n$, then there exists a similarity transformation T such that

$$\bar{A} = TAT^{-1}, \quad \bar{B} = TB, \quad \bar{C} = CT^{-1}$$

The transformed system attains the so-called staircase form with the uncontrollable modes, if any, in the upper left-hand corner.

$$\bar{A} = \begin{bmatrix} A_{uc} & 0 \\ A_{21} & A_c \end{bmatrix}, \quad \bar{B} = \begin{bmatrix} 0 \\ B_c \end{bmatrix},$$

where (A_c, B_c) is controllable and

$$G(s) = C_c(sI - A_c)^{-1}B_c = C(sI - A)^{-1}B$$

Similarly, the staircase form of observability is given by

$$\bar{A} = \begin{bmatrix} A_{uo} & A_{12} \\ 0 & A_o \end{bmatrix}, \quad \bar{C} = \begin{bmatrix} 0 & C_o \end{bmatrix}$$

There is a duality relationship between these two forms.

It is always desirable to properly scale the matrices and to get a balanced realization for use in model reduction. Model reduction in state space is achieved by first obtaining a vector **g**, which is the diagonal of the joint (controllability and observability) Grammian. The individual Grammians are given by

$$G_c = \int_0^\infty e^{\lambda A} BB^T e^{\lambda A^T} d\lambda \quad \text{and} \quad G_o = \int_0^\infty e^{\lambda A} C^T C e^{\lambda A^T} d\lambda$$

In the model reduction method, the original system

$$\dot{x} = Ax + Bu$$

$$y = Cx + Du$$

is rearranged in the following form:

$$\begin{bmatrix} \dot{x}_1 \\ \dot{x}_2 \end{bmatrix} = \begin{bmatrix} A_{11} & A_{12} \\ A_{21} & A_{22} \end{bmatrix} \begin{bmatrix} x_1 \\ x_2 \end{bmatrix} + \begin{bmatrix} B_1 \\ B_2 \end{bmatrix} u$$

$$y = \begin{bmatrix} C_1 & C_2 \end{bmatrix} \begin{bmatrix} x_1 \\ x_2 \end{bmatrix} + Du$$

Setting $\dot{x}_2 = 0$ and solving for x_1

$$\dot{x}_1 = \begin{bmatrix} A_{11} - A_{12}A_{22}^{-1}A_{21} \end{bmatrix} x_1 + \begin{bmatrix} B_1 - A_{12}A_{22}^{-1}B_2 \end{bmatrix} u$$

$$y = \begin{bmatrix} C_1 - C_2A_{22}^{-1}A_{21} \end{bmatrix} x_1 + \begin{bmatrix} D_1 - C_2A_{22}^{-1}B_2 \end{bmatrix} u$$

we have a reduced model that preserves static input–output characteristics. In this procedure, A_{22}^{-1} should exist.

6.1.3 Frequency-Domain Methods

The frequency-domain techniques start with a transfer function or transfer matrix description of the original system. The objective in this case is that the frequency-domain properties of the reduced-order equivalent match closely with those of the original system. These methods are not very much different in spirit from the optimization methods in the time domain outlined earlier.

The frequency response data required for these methods are practically more involved than the step response data in terms of experimental effort.

6.1.4 Time-Moment Matching

This method is based on determining a set of time moments of the impulse response function of the high-order system and matching them with those of the reduced model. The number of time moments matched depends on the desired order of the reduced model (Gibilaro and Lees, 1969). The more the number of time moments matched, the more accurate the low-order approximant. In the case of the continued fraction expansion, the degree of the numerator of the reduced-order model obtained is always either equal to or one less than the degree of the denominator.

6.1.5 Some Other Methods

Several researchers have studied the problem of approximating a high-order dynamical system by a low-order one, based on minimization of some norm. A balanced realization and truncation method was introduced by Moore (1981), where the controllability and observability Grammians are used to define measures of controllability and observability in certain directions of the state space. The Grammians are not invariant under coordinate transformations and it was shown that there exists a coordinated system in which the Grammians are equal and diagonal. The corresponding system representation is called *balanced*. A reduced-order model can be obtained from the balanced representation by deleting the least controllable and, therefore, least observable part. However, this technique in some cases is known to result in reduced-order models having a large steady-state error. This is due to the fact that balancing results in a mismatch of DC gains of the high-order model and the reduced-order model. Safonov Chiang (1989) gave an algorithm based on a Schur method for balanced-truncation model reduction. It is chosen so that a not necessarily balanced state-space realization of the Moore reduced model can be computed directly without balancing via projections defined in terms of arbitrary base for the left and right eigenspaces, associated with the *large* eigenvalues of the product of the (G_o) reacheability and (G_c) controllability Grammians. Two specific methods for computing these bases are proposed: one based on the ordered Schur decomposition of product (G_c) controllability and (G_o) reacheability and the other based on the Cholesky factors of controllability and reacheability.

In the Schur method (Safanov and Chiang, 1988), one computes a kth-order reduced model

$$\mathbf{G}_r(s) = \mathbf{C}_r(s\mathbf{I} - \mathbf{A}_r)^{-1}\mathbf{B}_r + \mathbf{D}_r,$$

such that the infinity norm of the error

$$\left\| \mathbf{G}(j\omega) - \mathbf{G}_r(j\omega) \right\|_\infty \leq 2\sum_{i=k+1}^{n} \sigma_i,$$

where σ_i are Hankel singular values given by $\sigma = \sqrt{\mathrm{eig}(\mathbf{G}_c * \mathbf{G}_o)}$

For unstable $\mathbf{G}(s)$, the algorithm works by first splitting $\mathbf{G}(s)$ into a sum of stable and unstable parts, that is, the square roots of eigenvalues of their reachability and observability Grammians:

$$\mathbf{G}_c = \int_0^\infty e^{\lambda A} \mathbf{B}\mathbf{B}^T e^{\lambda A^T} d\lambda \quad \text{and} \quad \mathbf{G}_o = \int_0^\infty e^{\lambda A} \mathbf{C}^T \mathbf{C} e^{\lambda A^T} d\lambda$$

Based on the Schur decomposition of $\mathbf{G}_c\mathbf{G}_o$ for computing orthonormal bases for the right eigenspaces of $\mathbf{G}_c\mathbf{G}_o$ associated with the dominant eigenvalues ($\sigma_1^2 \ldots \sigma_k^2$) and small eigenvalues ($\sigma_{k+1}^2 \ldots \sigma_n^2$), and also for the left eigenspace basis, *projection matrices* are developed to the reduced model. This method makes maximal use of orthogonal transformations and tends to ensure that the projection matrices are better conditioned when the system has some modes, which are much more observable than the controllable and/or vice versa.

6.2 Method of Approach for Reduction to First- or Second-Order Plus Dead-Time Forms

The present problem of model reduction may be stated as follows.

Given the step response $h_o(t)$ of a large original linear time-invariant type zero asymptotically stable single-input single-output system, find the transfer function $G_r(\mathbf{p}, s)$ of a reduced model G_r such that

$$J = \int_0^\infty \left[h_o(t) - L^{-1}\left\{ s^{-1}G_r(\mathbf{p}, s) \right\} \right]^2 dt$$

is minimized subject to

$$h_o(\infty) = G_r(\mathbf{p}, 0)$$

$$p_i > 0$$

for some specified i (parameters related to poles), and

$$p_n \geq 0$$

where $\mathbf{p}$ is an n vector of parameters of $G_r(\mathbf{p}, s)$ and the nth element of p corresponds to a time delay.

In the computation of J, the common practice is to get the Z-transform $G_r(\mathbf{q}, z)$ of $G_r(\mathbf{p}, s)$ and then use the related difference equation to compute the step response of the reduced model as

$$h_r(kT) = Z^{-1}\{G_r(\mathbf{q}, z)\}, \quad k = 0,1,2,\ldots,$$

where T is the sampling time. This procedure implies a rounding error when p_n/T is not an integer. In discretizing the continuous time model, zero-order hold assumption is also required. Of course, this assumption is valid for the step input case. However, fractional delay may sometimes render the discrete time transfer function nonminimum phase. Furthermore, in general, when no information is available regarding input signals between the sampling instants, it is not possible to relate it to a continuous time version.

In the present method, in view of the simplicity of the forms chosen for $G_r(\mathbf{p}, s)$, namely,

1. $G_r(\mathbf{p}, s) = Ke^{-\theta s}/(1 + \tau s)$ (first order plus delay) or
2. $G_r(\mathbf{p}, s) = K(1 + \tau_o s)e^{-\theta s}/(1 + \tau_1 s)(1 + \tau_2 s)$ (second order plus delay)

$h_r(t)$ may be analytically provided avoiding the sampling process on the model. The continuous time function corresponding to the delay free portion of the chosen model can be shifted as desired and then sampled, to get the response of the continuous time model with delay exactly at the required sampling instants for comparison with the given system response $h_o(t)$.

Next, minimization of J can be performed with the help of any standard routine. If an unconstrained minimization routing is available, the constraints can be applied externally. General routines for the time response of linear systems with a rigid vector of time instants should be avoided, because the use of a rigid time instant vector will imply the same problem with fractional delays as mentioned earlier.

6.3 The Case of an 18-Stage Multi-Stage Flash Desalination Plant Model

MSF plant models are very large in size and complexity. The nonlinear plant model is linearized at an operating point choosing six manipulated variables (inputs) and six controlled variables (outputs) into a standard state-space form as shown in the previous chapter. The matrices **A, B, C,** and **D** with the usual notation are obtained. The steady-state or DC gain matrix is computed as $G(0) = -C A^{-1}B$ easily when the matrix **A** is non-singular, that is, if the integrating loops of level control are closed and not considered in the input–output description. The minimal realization algorithm could remove only two redundant states out of the total of 155. Standard system theoretic methods of model reduction could not reduce this minimal model to a tractably simple lower order form, since the system states (153 in number) are uniformly scattered, making the elimination procedure unsuccessful even with heavy tolerances. The Schur algorithm could reduce the model to kth order; for example, taking $k = 6$ (to compute a sixth-order reduced model), it removes 149 states out of 155. It has a better approximation property at low frequencies and fair preservation of the steady-state (DC) gain. This method delivers an error bound (infinity norm) on the unmodeled dynamics due to model reduction and this will be useful in robust control. In this effort, robust control toolbox MATLAB® is employed.

6.4 Application of the Present Method

The matrices **A, B, C,** and **D** are first used to compute the DC gain matrix $G(0)$. The step response of the original plant model is obtained by using the routine *step* of MATLAB at an adequate number of points in time until the response settles in steady state. This data is used as a reference. The algorithm is initiated with a parameter vector in the reduced model. The step response of the reduced model for this parameter vector is computed and used in the computation of the objective function J. The MATLAB routine FMINS is employed with external constraints to minimize J. Both the first- and second-order forms with delay are fitted, and the results are shown for G_{16} in Figure 6.1a through 6.1c. When the step response of the original unreduced model has no overshoot, fitting with a first-order model with a delay is surprisingly good. When an overshoot exists as is seen in some cases, the first-order model is unable to account for the same. It is the second order plus delay model, which fits such situations extremely well. In the present reduction procedure, the steady-state gain is preserved during reduction, which also simplifies the minimization procedure in terms of the number of variables with respect to which the search for the minimum should be conducted. In some cases where the *first order with delay* structure was fitted well, the *second order with delay* structure could hardly improve the fit any further. The corresponding frequency response functions are

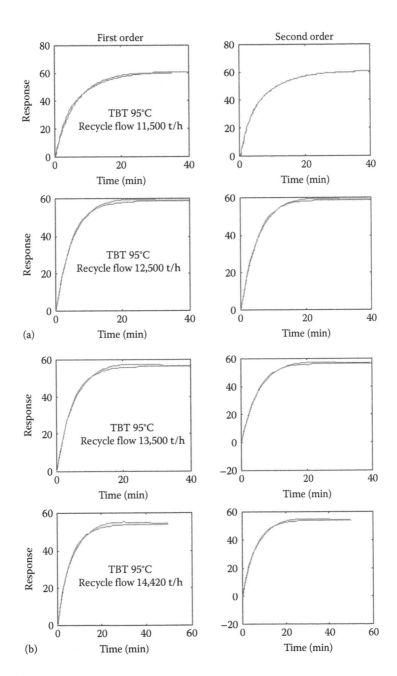

FIGURE 6.1
(a–f) Step response of original model and reduced model of $G(1, 6)$. *(Continued)*

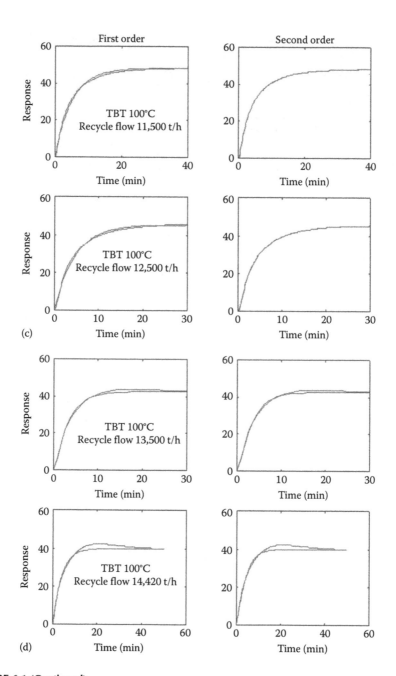

FIGURE 6.1 (*Continued*)
(a–f) Step response of original model and reduced model of $G(1, 6)$.   (*Continued*)

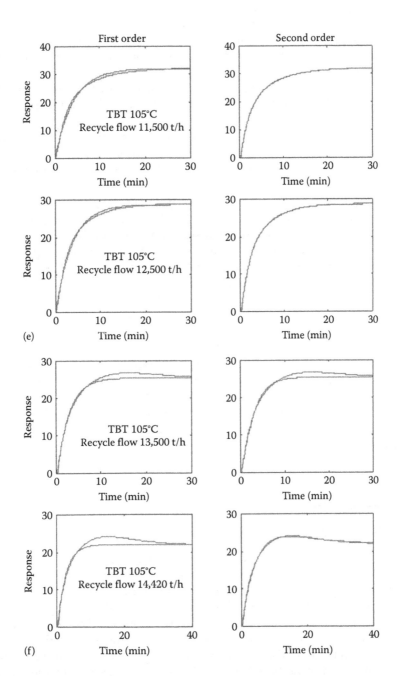

FIGURE 6.1 (*Continued*)
(a–f) Step response of original model and reduced model of $G(1, 6)$.

shown in Figure 6.A.1a through l that shows the step response plots of the unreduced plant model in the 12 cases of study.

Notice that only four elements of the table in the first row, that is, the second, third, fifth, and sixth, need reduction. There are already some first-order elements.

The fourth element in the first row (y_1/u_4) (y_1/u_1) whose magnitude is negligible relative to the others is parasitic and therefore need not be considered.

Figure 6.A.2a through l gives an idea of the approximation also in the frequency domain. The frequency response characteristics of the reduced models are seen to match the original model response extremely well in the frequency band of practical interest. This is not surprising in view of the excellent matching noticed already in the response. Section 6.A.3 in the Appendix shows the reduced TFMs at the operating conditions considered earlier. Here we will consider only the top brine temperature (TBT) control that requires $G(1,6)$ in the TFM for which the step response functions of the full model are compared with the reduced first- and second-order plus delay versions in different cases.

The TFMs with the reduced models are shown at two orders of approximation. Wherever relevant, both first- and second-order models are included in separate TFMs. When a first order plus delay model is inserted, which is for G_{15} and/or G_{16}, the TFM is labeled as *first order*. Similarly, when a second-order plus delay model is inserted either for G_{15} and/or G_{16}, it is labeled as *second order*. Thus, we have the model of the MSF plant at two levels of simplicity for further use according to the choice in design. Since the insertions for G_{15} and/or G_{16}, in the TFMs have been made after discarding the cases of poor fit, any one of the versions, namely, *first order* or *second order*, will be satisfactory.

6.5 Discussion and Conclusion

A numerically efficient procedure to get an optimally reduced model on the basis of step response data for large and otherwise not easily reducible models is presented. The procedure can be modified to suit data from the response due to finite duration pulses. The analytical expressions for the model response in the chosen form may be obtained by a suitable combination of step response expressions incorporating sign and delay. The procedure can be used also with experimentally obtained data. Many modifications of the basic method discussed here are possible suiting the available tools for optimization. The model in the reduced form is now suitable for control system design considerations.

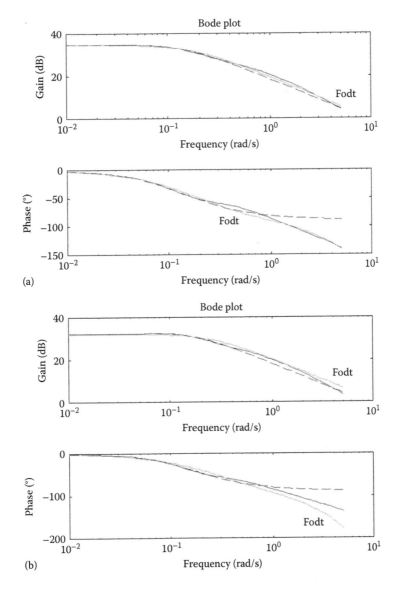

FIGURE 6.A.2
(a) $G_{95,14,420}(1,6)$—Case 1, (b) $G_{100,14,420}(1,6)$—Case 2 (- - - - SODT; ———— Full model).   (*Continued*)

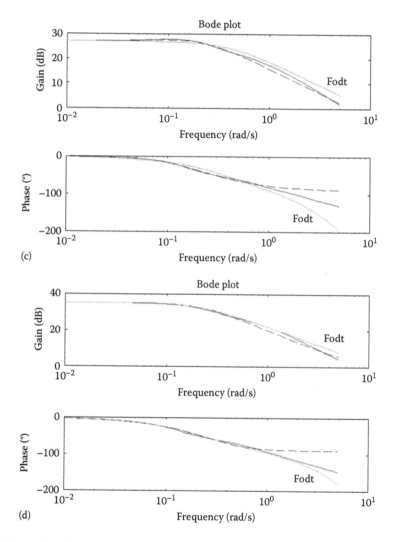

FIGURE 6.A.2 (Continued)
(c) $G_{105,14,420}(1,6)$—Case 3, (d) $G_{95,13.500}(1,6)$—Case 4 (- - - - - SODT; ——— Full model).
(*Continued*)

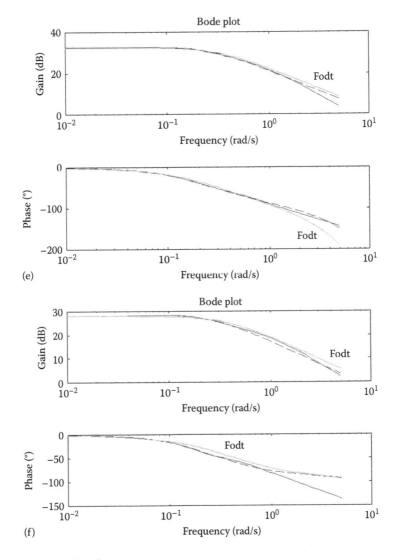

FIGURE 6.A.2 (Continued)
(e) $G_{100,13.500}(1,6)$—Case 5, (f) $G_{105,13.500}(1,6)$—Case 6 (- - - - - SODT; ———— Full model).

(Continued)

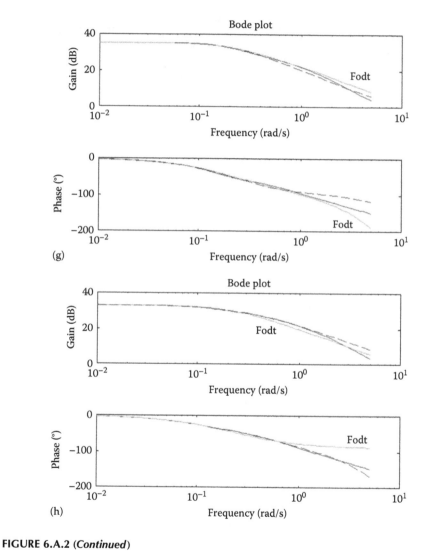

FIGURE 6.A.2 (*Continued*)
(g) $G_{95,12.500}(1,6)$—Case 7, (h) $G_{100,12.500}(1,6)$—Case 8 (- - - - SODT; ——— Full model).

(*Continued*)

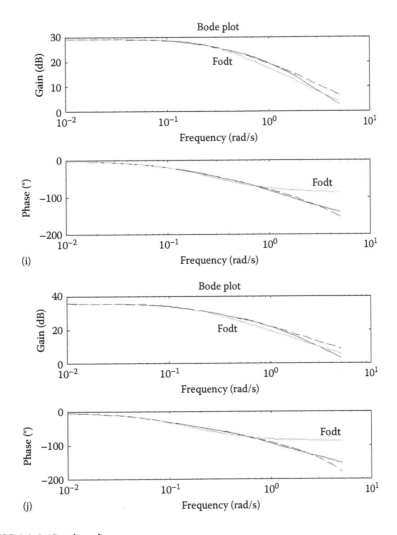

FIGURE 6.A.2 (*Continued*)
(i) $G_{105,12.500}(1,6)$—Case 9, (j) $G_{95,11.500}(1,6)$—Case 10 (- - - - - SODT; ———— Full model).

(*Continued*)

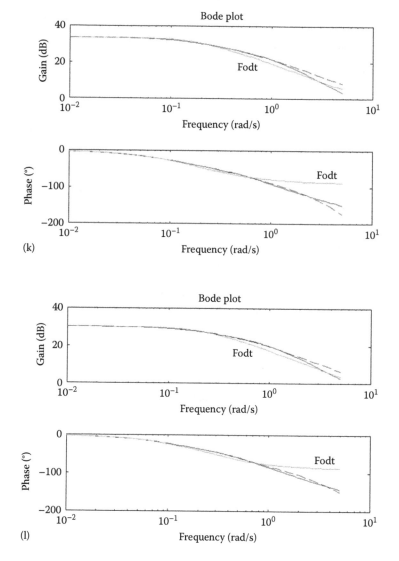

FIGURE 6.A.2 (*Continued*)
(k) $G_{100,11.500}(1,6)$—Case 11, and (l) $G_{105,11.500}(1,6)$—Case 12 (- - - - SODT; ——— Full model).

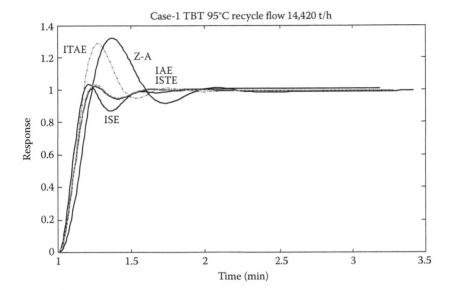

FIGURE 7.5
Step response of the optimally tuned TBT control loop.

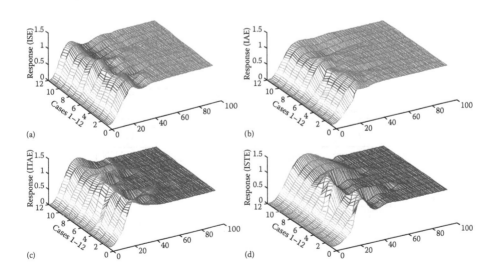

FIGURE 7.6
Step response of the optimally tuned systems in the operating space: (a) responses for optimal PID settings using ISE criterion, (b) responses for optimal PID settings using IAE criterion, (c) responses for optimal PID settings using ITAE criterion, and (d) responses for optimal PID settings using ISTE criterion.

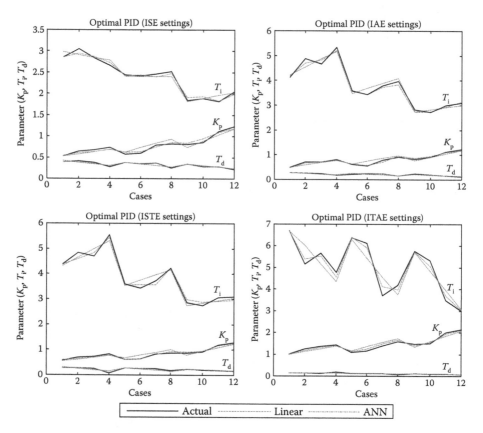

FIGURE 8.5
Actual and predicted outputs.

6.A Appendix Assessment of Reduced Models

6.A.1 Assessment in Time Domain

Step response matrices (Figure 6.A.1a through l) of the unreduced TFMs in different cases.

6.A.2 Comparison of Two Versions of Reduced Model

Comparison of the first and second order plus delay approximations with the full models in the frequency domain (Figure 6.A.2a through l).

6.A.3 Reduced Transfer Function Matrices for $G_{TBT, REC}$ (1,5) and $G_{TBT, REC}$ (1,6)

Case 1: TBT 95°C and recycle flow 14,420 t/h
First order plus delay approximation

$$
\begin{bmatrix}
0 & \dfrac{-2.1884e^{-2.55s}}{(1+8.10s)} & \dfrac{-82.032e^{-0.22s}}{(1+1.52s)} & 0 & \dfrac{-2.8\,e^{-6.5s}}{(1+10.98s)} & \dfrac{54\,e^{-0.187s}}{(1+5.76s)} \\
61.7 & 176.2433 & 0 & 0 & 183.3005 & 0 \\
0 & 176.2433 & 0 & \dfrac{96.7}{1+0.1s} & 183.3005 & 0 \\
0 & 0 & 541.0021 & 0 & 0 & 0 \\
0 & 0 & 0 & \dfrac{3.496}{1+0.1s} & 0 & 0 \\
0 & 176.2433 & 0 & 0 & 0 & 0
\end{bmatrix}
$$

Second order plus delay approximation

$$
\begin{bmatrix}
0 & \dfrac{-2.1884e^{-1.4676s}(1+28.8135s)}{(1+17.522s)(1+17.522s)} & \dfrac{-82.032e^{-0.0333s}(1+24.878s)}{(1+8.576s)(1+8.460s)} & 0 & \dfrac{-2.8e^{-6.5s}}{(1+10.98s)} & \dfrac{54(1+20.32s)}{(1+18.3s)(1+7.2s)} \\
61.7 & 176.2433 & 0 & 0 & 183.3005 & 0 \\
0 & 176.2433 & 0 & \dfrac{96.7}{1+0.1s} & 183.3005 & 0 \\
0 & 0 & 541.0021 & 0 & 0 & 0 \\
0 & 0 & 0 & \dfrac{3.496}{1+0.1s} & 0 & 0 \\
0 & 176.2433 & 0 & 0 & 0 & 0
\end{bmatrix}
$$

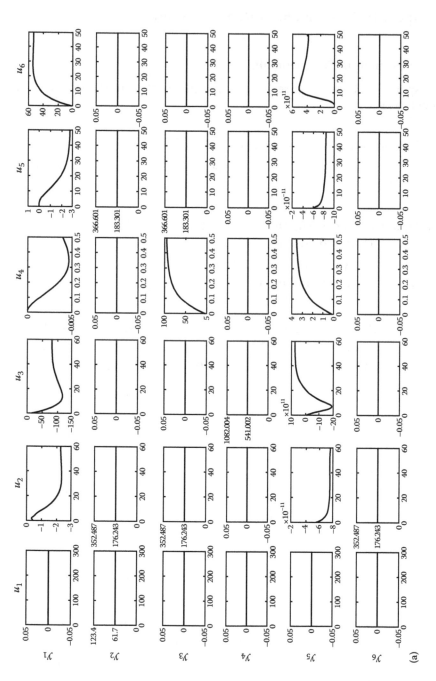

(Continued)

FIGURE 6.A.1
(a) TBT = 95°C; recycle flow = 14,420 t/h.

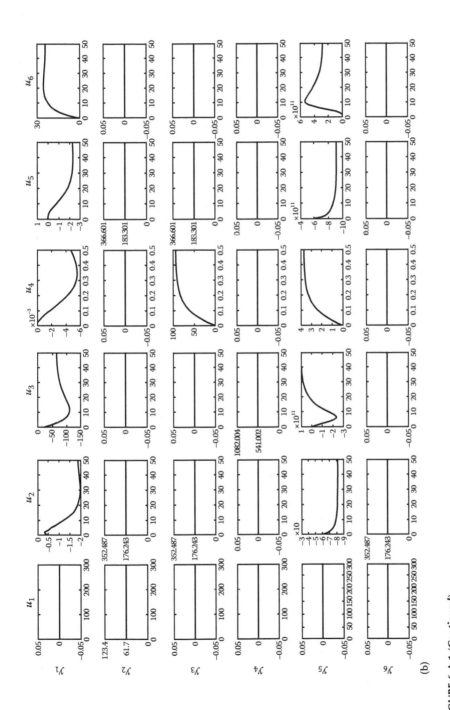

(Continued)

FIGURE 6.A.1 (Continued)
(b) TBT = 100°C; recycle flow = 14,420 t/h.

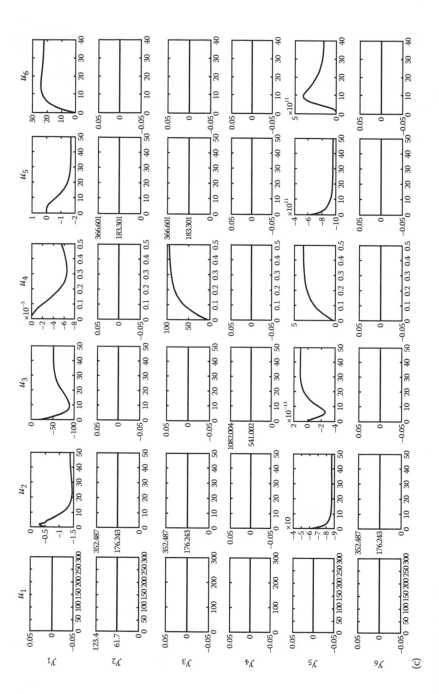

(Continued)

FIGURE 6.A.1 (Continued)
(c) TBT = 105°C; recycle flow = 14,420 t/h.

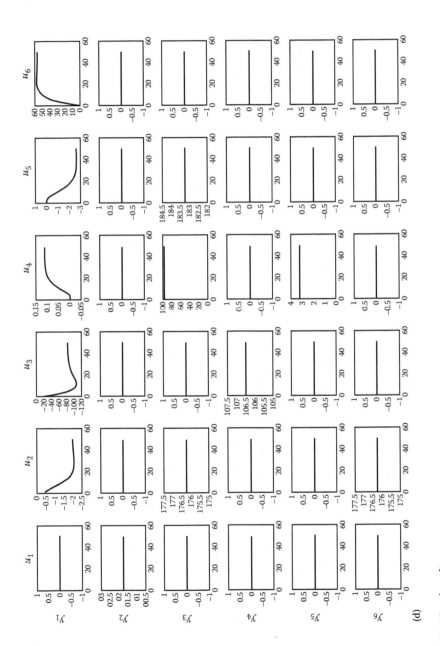

FIGURE 6.A.1 (Continued)
(d) TBT = 95°C; recycle flow = 13,520 t/h.

(Continued)

Multi-Stage Flash Desalination

(Continued)

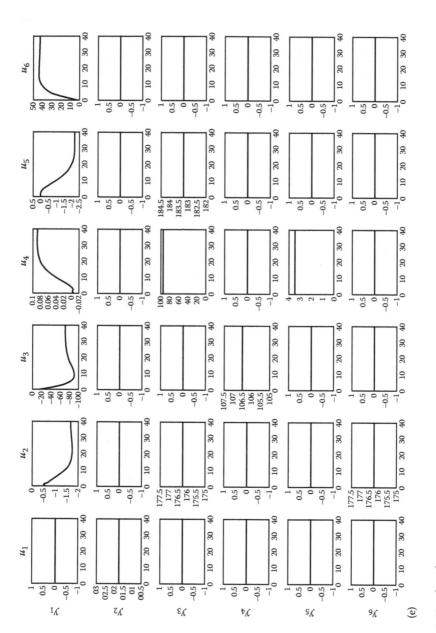

FIGURE 6.A.1 (Continued)

(e) TBT = 100°C; recycle flow = 13,520 t/h.

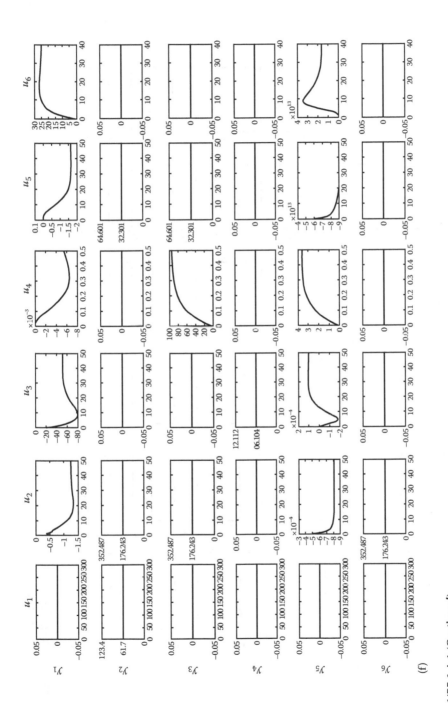

(Continued)

FIGURE 6.A.1 (Continued)
(f) TBT = 105°C; recycle flow = 13,520 t/h.

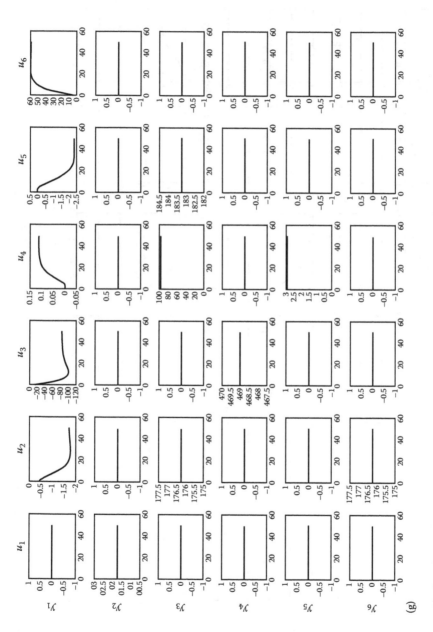

(Continued)

FIGURE 6.A.1 (*Continued*)
(g) TBT = 95°C; recycle flow = 12,520 t/h.

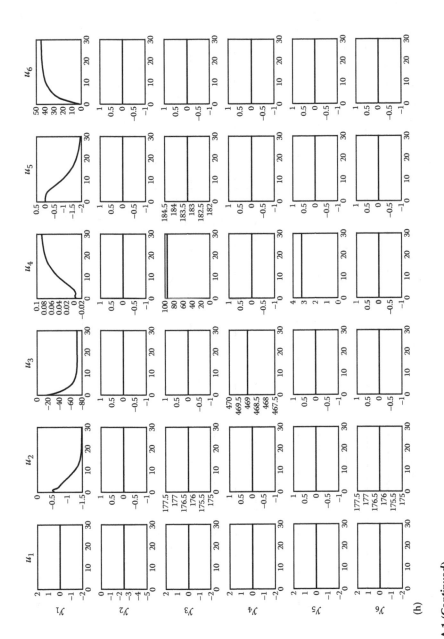

FIGURE 6.A.1 (Continued)
(h) TBT = 100°C; recycle flow = 12,520 t/h.

(Continued)

(Continued)

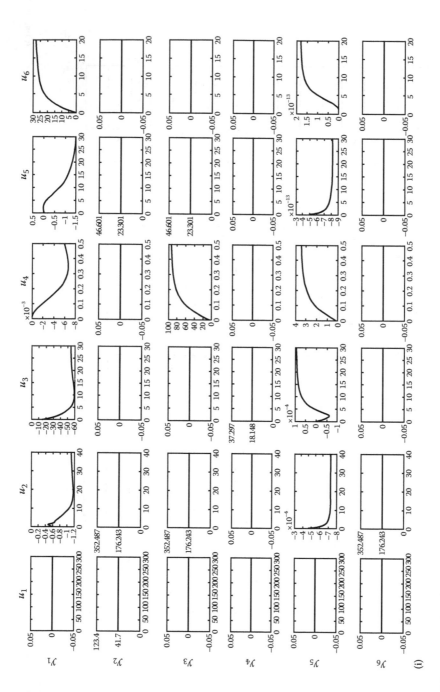

FIGURE 6.A.1 (Continued)

(i) TBT = 105°C; recycle flow = 12,520 t/h.

(*Continued*)

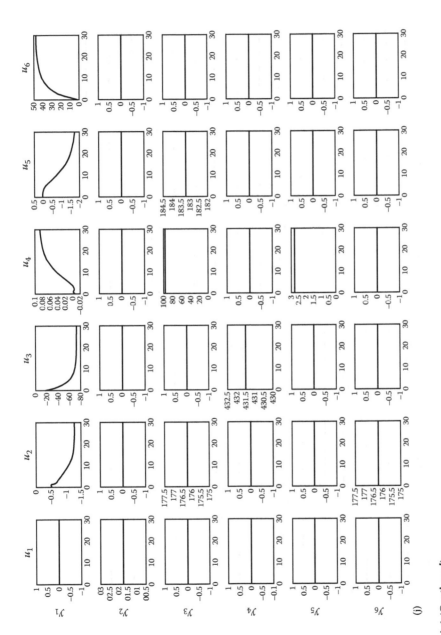

FIGURE 6.A.1 (Continued)
(j) TBT = 95°C; recycle flow = 11,520 t/h.

(Continued)

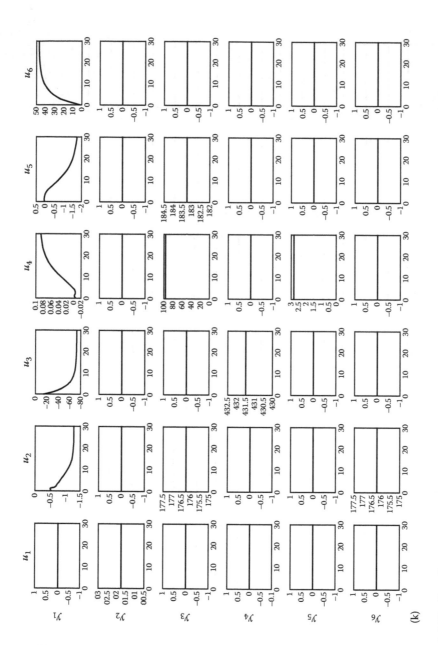

FIGURE 6.A.1 (Continued)
(k) TBT = 100°C; recycle flow = 11,520 t/h.

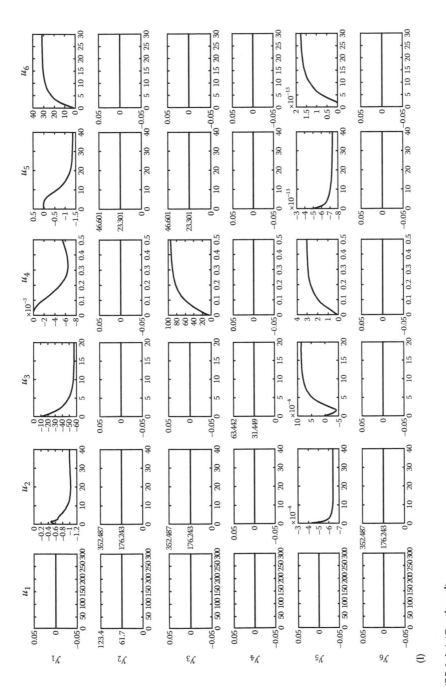

FIGURE 6.A.1 (Continued)
(I) TBT = 105°C; recycle flow = 11,520 t/h.

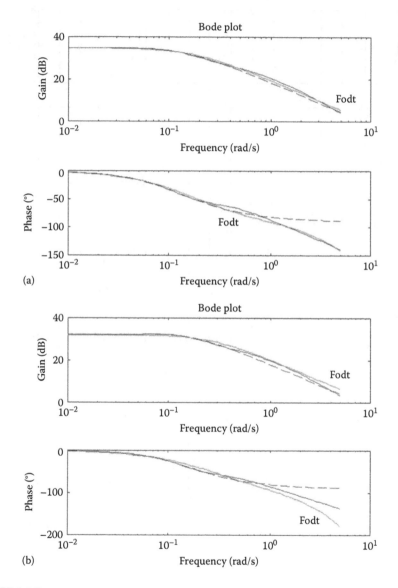

FIGURE 6.A.2
(See color insert.) (a) $G_{95,14,420}(1,6)$—Case 1, (b) $G_{100,14,420}(1,6)$—Case 2 (- - - - - SODT; ———— Full model). *(Continued)*

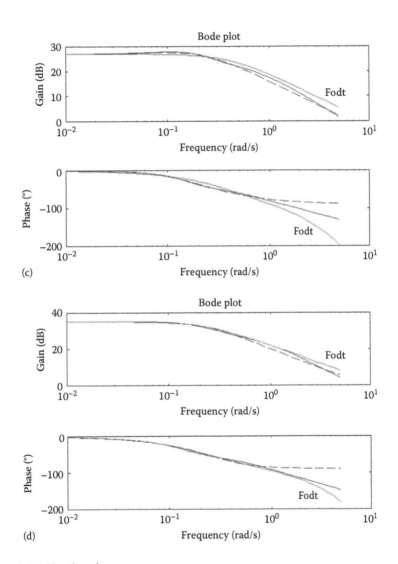

FIGURE 6.A.2 (*Continued*)
(See color insert.) (c) $G_{105,14,420}(1,6)$—Case 3, (d) $G_{95,13.500}(1,6)$—Case 4 (- - - - SODT; ———— Full model). (*Continued*)

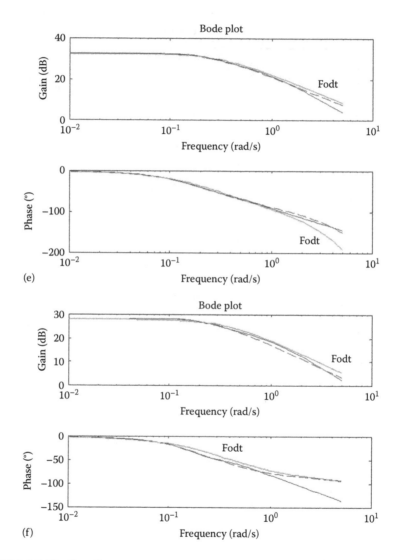

FIGURE 6.A.2 (*Continued*)
(See color insert.) (e) $G_{100,13.500}(1,6)$—Case 5, (f) $G_{105,13.500}(1,6)$—Case 6 (- - - - - SODT; ——— Full model). (*Continued*)

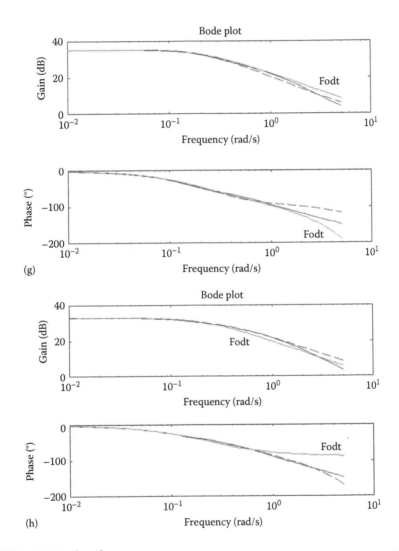

FIGURE 6.A.2 (*Continued*)
(See color insert.) (g) $G_{95,12.500}(1,6)$—Case 7, (h) $G_{100,12.500}(1,6)$—Case 8 (- - - - SODT; ——— Full model). (*Continued*)

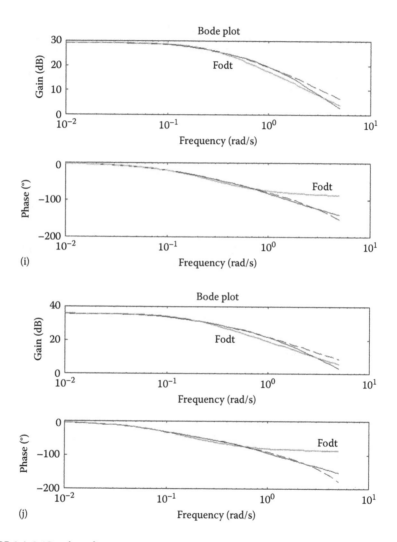

FIGURE 6.A.2 (*Continued*)
(See color insert.) (i) $G_{105,12.500}(1,6)$—Case 9, (j) $G_{95,11.500}(1,6)$—Case 10 (- - - - - SODT; ———— Full model).   (*Continued*)

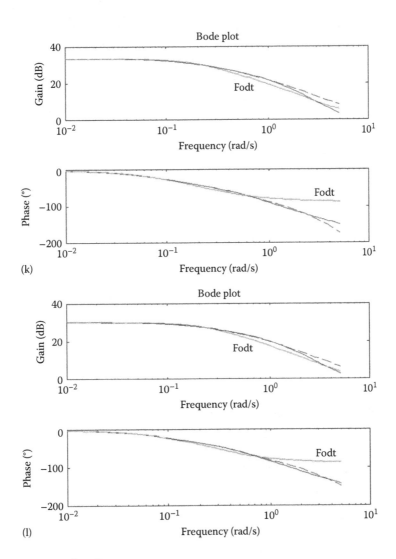

FIGURE 6.A.2 (*Continued*)
(**See color insert.**) (k) $G_{100,11.500}(1,6)$—Case 11, and (l) $G_{105,11.500}(1,6)$—Case 12 (- - - - - SODT; ———
Full model).

Case 2: TBT 100°C and recycle flow 14,420 t/h
First order plus delay approximation

$$
\begin{bmatrix}
0 & \dfrac{-1.83e^{-1.35s}}{(1+6.6s)} & \dfrac{-68.76e^{-0.14s}}{(1+1.1s)} & 0 & \dfrac{-2.31e^{-6.15s}}{(1+8.8s)} & \dfrac{39.85\,e^{-0.32s}}{(1+3.79s)} \\
61.7 & 176.2433 & 0 & 0 & 183.3005 & 0 \\
0 & 176.2433 & 0 & \dfrac{96}{1+0.1s} & 183.3005 & 0 \\
0 & 0 & 541.0021 & 0 & 0 & 0 \\
0 & 0 & 0 & \dfrac{3.76}{1+0.1s} & 0 & 0 \\
0 & 176.2433 & 0 & 0 & 0 & 0
\end{bmatrix}
$$

Second order plus delay approximation

$$
\begin{bmatrix}
0 & \dfrac{-1.8275(1+28.2734s)}{(1+16.2508s)(1+16.4070s)} & \dfrac{-68.7636(1+25.4626s)}{(1+7.5466s)(1+7.5400s)} & 0 & \dfrac{-2.3144e^{-3.3593s}}{(1+5.6254s)(1+5.6254s)} & \dfrac{39.85(1+13.55s)}{(1+8.145s)(1+8.2565s)} \\
61.7 & 176.2433 & 0 & 0 & 183.3005 & 0 \\
0 & 176.2433 & 0 & \dfrac{96.7}{1+0.1s} & 183.3005 & 0 \\
0 & 0 & 541.0021 & 0 & 0 & 0 \\
0 & 0 & 0 & \dfrac{3.76}{1+0.1s} & 0 & 0 \\
0 & 176.2433 & 0 & 0 & 0 & 0
\end{bmatrix}
$$

Case 3: TBT 105°C and recycle flow 14,420 t/h
First order plus delay approximation

$$
\begin{bmatrix}
0 & \dfrac{-1.36e^{-0.0s}}{(1+5.49s)} & \dfrac{-51.56e^{-0.07s}}{(1+0.666s)} & 0 & \dfrac{-1.6959e^{-5.4286s}}{(1+7.4393s)} & \dfrac{22.05e^{-0.391s}}{(1+2.3586s)} \\
61.7 & 176.2433 & 0 & 0 & 183.3005 & 0 \\
0 & 176.2433 & 0 & \dfrac{96.7}{1+0.1s} & 183.3005 & 0 \\
0 & 0 & 541.0021 & 0 & 0 & 0 \\
0 & 0 & 0 & \dfrac{4.0395}{1+0.1s} & 0 & 0 \\
0 & 176.2433 & 0 & 0 & 0 & 0
\end{bmatrix}
$$

Second order plus delay approximation

$$
\begin{bmatrix}
0 & \dfrac{-1.3565e^{-0.1258s}(1+18.5s)}{(1+11.0583s)(1+10.9493s)} & \dfrac{-51.5614e^{-0.1897s}(1+24.3265s)}{(1+6.3047s)(1+6.3047s)} & 0 & \dfrac{-1.6959(1-2.4445s)}{(1+5.27s)(1+5.27s)} & \dfrac{22.05(1+13.1419s)}{(1+9.3851s)(1+4.797s)} \\
61.7 & 176.2433 & 0 & 0 & 183.3005 & 0 \\
0 & 176.2433 & 0 & \dfrac{96.7}{1+0.1s} & 183.3005 & 0 \\
0 & 0 & 541.0021 & 0 & 0 & 0 \\
0 & 0 & 0 & \dfrac{4.03}{1+0.1s} & 0 & 0 \\
0 & 176.2433 & 0 & 0 & 0 & 0
\end{bmatrix}
$$

Case 4: TBT 95°C and recycle flow 13,520 t/h
First order plus delay approximation

$$
\begin{bmatrix}
0 & \dfrac{-1.98e^{-0.0s}}{(1+7.07s)} & \dfrac{-83.53e^{-0.16s}}{(1+1.57s)} & 0 & \dfrac{-2.6e^{-5.85s}}{(1+7.91s)} & \dfrac{56.3e^{-0.323s}}{(1+4.42s)} \\
61.7 & 176.2433 & 0 & 0 & 183.3005 & 0 \\
0 & 176.2433 & 0 & \dfrac{96.7}{1+0.1s} & 183.3005 & 0 \\
0 & 0 & 506.486 & 0 & 0 & 0 \\
0 & 0 & 0 & \dfrac{3.22}{1+0.1s} & 0 & 0 \\
0 & 176.2433 & 0 & 0 & 0 & 0
\end{bmatrix}
$$

Second order plus delay approximation

$$
\begin{bmatrix}
0 & \dfrac{-1.98e^{-0.0s}(1+1.69s)}{(1+4.40s)(1+4.38s)} & \dfrac{-83.53e^{-0.0s}(1+15.3s)}{(1+6.92s)(1+5.86s)} & 0 & \dfrac{-83.53e^{-0.0s}(1+15.3s)}{(1+6.92s)(1+5.86s)} & \dfrac{56.3e^{-0.0s}(1+1.65s)}{(1+3.15s)(1+3.15s)} \\
61.7 & 176.2433 & 0 & 0 & 183.3005 & 0 \\
0 & 176.2433 & 0 & \dfrac{96.7}{1+0.1s} & 183.3005 & 0 \\
0 & 0 & 506.486 & 0 & 0 & 0 \\
0 & 0 & 0 & \dfrac{3.22}{1+0.1s} & 0 & 0 \\
0 & 176.2433 & 0 & 0 & 0 & 0
\end{bmatrix}
$$

Case 5: TBT 100°C and recycle flow 13,520 t/h
First order plus delay approximation

$$
\begin{bmatrix}
0 & \dfrac{-1.66e^{-0.0s}}{(1+4.95s)} & \dfrac{-70.4e^{-0.12s}}{(1+1.14s)} & 0 & \dfrac{-2.17e^{-5.4s}}{(1+6.45s)} & \dfrac{42.6e^{-0.365s}}{(1+3.18s)} \\
61.7 & 176.2433 & 0 & 0 & 183.3005 & 0 \\
0 & 176.2433 & 0 & \dfrac{96.7}{1+0.1s} & 183.3005 & 0 \\
0 & 0 & 506.486 & 0 & 0 & 0 \\
0 & 0 & 0 & \dfrac{3.47}{1+0.1s} & 0 & 0 \\
0 & 176.2433 & 0 & 0 & 0 & 0
\end{bmatrix}
$$

Second order plus delay approximation

$$
\begin{bmatrix}
0 & \dfrac{-1.66e^{-0.04s}(1+0.73s)}{(1+0.05s)(1+5.35s)} & \dfrac{-70.4e^{0.13s}(1+13.1s)}{(1+5.24s)(1+5.24s)} & 0 & \dfrac{-2.17e^{-3.3s}(1+0.0s)}{(1+4.15s)(1+4.15s)} & \dfrac{42.6e^{-0.22s}(1+11.95s)}{(1+10.54s)(1+4.16s)} \\
61.7 & 176.2433 & 0 & 0 & 183.3005 & 0 \\
0 & 176.2433 & 0 & \dfrac{96.7}{1+0.1s} & 183.3005 & 0 \\
0 & 0 & 506.486 & 0 & 0 & 0 \\
0 & 0 & 0 & \dfrac{3.47}{1+0.1s} & 0 & 0 \\
0 & 176.2433 & 0 & 0 & 0 & 0
\end{bmatrix}
$$

Case 6: TBT 105°C and recycle flow 13,500 t/h
First order plus delay approximation

$$
\begin{bmatrix}
0 & \dfrac{-1.245e^{-0.0s}}{(1+4.35s)} & \dfrac{-53.33e^{-0.05s}}{(1+0.85s)} & 0 & \dfrac{-1.6077e^{-5.55s}}{(1+6.5611s)} & \dfrac{25.34\,e^{-0.028s}}{(1+2.697s)} \\
61.7 & 176.2433 & 0 & 0 & 183.3005 & 0 \\
0 & 176.2433 & 0 & \dfrac{96.7}{1+0.1s} & 183.3005 & 0 \\
0 & 0 & 506.456 & 0 & 0 & 0 \\
0 & 0 & 0 & \dfrac{3.7235}{1+0.1s} & 0 & 0 \\
0 & 176.2433 & 0 & 0 & 0 & 0
\end{bmatrix}
$$

Second order plus delay approximation

$$
\begin{bmatrix}
0 & \dfrac{-1.245(1+15.5514s)}{(1+8.1591s)(1+10.2787s)} & \dfrac{-53.33e^{-0.0s}(1+17.42s)}{(1+5.59s)^2} & 0 & \dfrac{-1.6077e^{-3.4711s}}{(1+4.1986s)(1+4.1987s)} & \dfrac{25.338se^{-0.02s}(1+11.91s)}{(1+9.71s)(1+4.23s)} \\
61.7 & 176.2433 & 0 & 0 & 183.3005 & 0 \\
0 & 176.2433 & 0 & \dfrac{96.7}{1+0.1s} & 183.3005 & 0 \\
0 & 0 & 506.486 & 0 & 0 & 0 \\
0 & 0 & 0 & \dfrac{3.7235}{1+0.1s} & 0 & 0 \\
0 & 176.2433 & 0 & 0 & 0 & 0
\end{bmatrix}
$$

Case 7: TBT 95°C and recycle flow 12,500 t/h
First order plus delay approximation

$$
\begin{bmatrix}
0 & \dfrac{-1.74e^{-0.0s}}{(1+6.35s)} & \dfrac{-84.93e^{0.17s}}{(1+1.74s)} & 0 & \dfrac{-2.38e^{-6.2s}}{(1+7.6s)} & \dfrac{58.6e^{-0.362s}}{(1+4.45s)} \\
61.7 & 176.2433 & 0 & 0 & 183.3005 & 0 \\
0 & 176.2433 & 0 & \dfrac{96.7}{1+0.1s} & 183.3005 & 0 \\
0 & 0 & 468.968 & 0 & 0 & 0 \\
0 & 0 & 0 & \dfrac{2.929}{1+0.1s} & 0 & 0 \\
0 & 176.2433 & 0 & 0 & 0 & 0
\end{bmatrix}
$$

Second order plus delay approximation

$$
\begin{bmatrix}
0 & \dfrac{-1.74e^{-0.0s}(1+3.25s)}{(1+4.79s)(1+4.78s)} & \dfrac{-84.93e^{-0.0s}(1+13.6s)}{(1+5.95s)(1+6.38s)} & 0 & \dfrac{-2.38e^{-3.69s}(1+0s)}{(1+4.91s)(1+4.91s)} & \dfrac{58.6e^{-0.098s}(1+1.7s)}{(1+3.2s)(1+3.2s)} \\
61.7 & 176.2433 & 0 & 0 & 183.3005 & 0 \\
0 & 176.2433 & 0 & \dfrac{96.7}{1+0.1s} & 183.3005 & 0 \\
0 & 0 & 468.968 & 0 & 0 & 0 \\
0 & 0 & 0 & \dfrac{2.929}{1+0.1s} & 0 & 0 \\
0 & 176.2433 & 0 & 0 & 0 & 0
\end{bmatrix}
$$

Case 8: TBT 100°C and recycle flow 12,500 t/h
First order plus delay approximation

$$\begin{bmatrix} 0 & \dfrac{-1.47e^{-0.0s}}{(1+4.72s)} & \dfrac{-71.99e^{-0.0s}}{(1+2.34s)} & 0 & \dfrac{-2.0e^{-5.14s}}{(1+7.5s)} & \dfrac{45.5e^{-0.0005s}}{(1+4.68s)} \\[2ex] 61.7 & 176.2433 & 0 & 0 & 183.3005 & 0 \\[2ex] 0 & 176.2433 & 0 & \dfrac{96.7}{1+0.1s} & 183.3005 & 0 \\[2ex] 0 & 0 & 468.968 & 0 & 0 & 0 \\[2ex] 0 & 0 & 0 & \dfrac{3.16}{1+0.1s} & 0 & 0 \\[2ex] 0 & 176.2433 & 0 & 0 & 0 & 0 \end{bmatrix}$$

Second order plus delay approximation

$$\begin{bmatrix} 0 & \dfrac{-1.47e^{-0.12s}(1+1.32s)}{(1+5.88s)(1+0.003s)} & \dfrac{-71.99e^{-0.02s}(1+0.53s)}{(1+0.15s)(1+2.72s)} & 0 & \dfrac{-2.0e^{-2.9s}(1+0.0s)}{(1+6.13s)(1+3.35s)} & \dfrac{45.5e^{-0.29s}(1+3.75s)}{(1+2.05s)(1+6.32s)} \\[2ex] 61.7 & 176.2433 & 0 & 0 & 183.3005 & 0 \\[2ex] 0 & 176.2433 & 0 & \dfrac{96.7}{1+0.1s} & 183.3005 & 0 \\[2ex] 0 & 0 & 468.968 & 0 & 0 & 0 \\[2ex] 0 & 0 & 0 & \dfrac{3.16}{1+0.1s} & 0 & 0 \\[2ex] 0 & 176.2433 & 0 & 0 & 0 & 0 \end{bmatrix}$$

Case 9: TBT 105°C and recycle flow 12,500 t/h
First order plus delay approximation

$$\begin{bmatrix} 0 & \dfrac{-1.116e^{-0.0s}}{(1+3.39s)} & \dfrac{-55.129e^{-0.0s}}{(1+1.52s)} & 0 & \dfrac{-1.5e^{-5.0549s}}{(1+6.987s)} & \dfrac{28.785}{(1+3.73s)} \\[2ex] 61.7 & 176.2433 & 0 & 0 & 183.3005 & 0 \\[2ex] 0 & 176.2433 & 0 & \dfrac{96.7}{1+0.1s} & 183.3005 & 0 \\[2ex] 0 & 0 & 468.97 & 0 & 0 & 0 \\[2ex] 0 & 0 & 0 & \dfrac{3.7235}{1+0.1s} & 0 & 0 \\[2ex] 0 & 176.2433 & 0 & 0 & 0 & 0 \end{bmatrix}$$

Second order plus delay approximation

$$
\begin{bmatrix}
0 & \dfrac{-1.1162(1 + 13.524s)}{(1 + 11.531s)(1 + 5.09s)} & \dfrac{-55.129(1 + 11.6778s)}{(1 + 10.0155s)(1 + 2.4639s)} & 0 & \dfrac{-1.5e^{-2.9552s}}{(1 + 4.3816s)(1 + 4.3816s)} & \dfrac{28.785e^{-0.2417s}(1 + 3.26s)}{(1 + 5.2s)(1 + 1.7s)} \\[3mm]
61.7 & 176.2433 & 0 & 0 & 183.3005 & 0 \\[2mm]
0 & 176.2433 & 0 & \dfrac{96.7}{1 + 0.1s} & 183.3005 & 0 \\[3mm]
0 & 0 & 468.97 & 0 & 0 & 0 \\[2mm]
0 & 0 & 0 & \dfrac{3.3833}{1 + 0.1s} & 0 & 0 \\[3mm]
0 & 176.2433 & 0 & 0 & 0 & 0
\end{bmatrix}
$$

Case 10: TBT 95°C and recycle flow 11,500 t/h
First order plus delay approximation

$$
\begin{bmatrix}
0 & \dfrac{-1.51e^{-0s}}{(1 + 6.44s)} & \dfrac{-86.2e^{-0.49s}}{(1 + 4.33s)} & 0 & \dfrac{-2.16e^{-5.6s}}{(1 + 9.48s)} & \dfrac{60.9e^{-0.0001s}}{(1 + 6.42s)} \\[3mm]
61.7 & 176.2433 & 0 & 0 & 183.3005 & 0 \\[2mm]
0 & 176.2433 & 0 & \dfrac{96.7}{1 + 0.1s} & 183.3005 & 0 \\[3mm]
0 & 0 & 431.450 & 0 & 0 & 0 \\[2mm]
0 & 0 & 0 & \dfrac{2.64}{1 + 0.1s} & 0 & 0 \\[3mm]
0 & 176.2433 & 0 & 0 & 0 & 0
\end{bmatrix}
$$

Second order plus delay approximation

$$
\begin{bmatrix}
0 & \dfrac{-1.51e^{-0.01s}(1 + 1.66s)}{(1 + 8.09s)(1 + 0.01s)} & \dfrac{-86.2e^{-0s}(1 + 52s)}{(1 + 0.594s)(1 + 4.89s)} & 0 & \dfrac{-2.16e^{-3.25s}}{(1 + 8.15s)(1 + 3.4s)} & \dfrac{60.9e^{-0.323s}(1 + 4.2s)}{(1 + 2.28s)(1 + 8.29s)} \\[3mm]
61.7 & 176.2433 & 0 & 0 & 183.3005 & 0 \\[2mm]
0 & 176.2433 & 0 & \dfrac{96.7}{1 + 0.1s} & 183.3005 & 0 \\[3mm]
0 & 0 & 431.450 & 0 & 0 & 0 \\[2mm]
0 & 0 & 0 & \dfrac{2.64}{1 + 0.1s} & 0 & 0 \\[3mm]
0 & 176.2433 & 0 & 0 & 0 & 0
\end{bmatrix}
$$

Case 11: TBT 100°C and recycle flow 11,500 t/h
First order plus delay approximation

$$
\begin{bmatrix}
0 & \dfrac{-1.28e^{-0.0s}}{(1+3.96s)} & \dfrac{-73.4e^{0.41s}}{(1+3.32s)} & 0 & \dfrac{-1.82e^{-5.31s}}{(1+8.27s)} & \dfrac{48.24e^{-0.0004s}}{(1+5.11s)} \\
61.7 & 176.2433 & 0 & 0 & 183.3005 & 0 \\
0 & 176.2433 & 0 & \dfrac{96.7}{1+0.1s} & 183.3005 & 0 \\
0 & 0 & 431.450 & 0 & 0 & 0 \\
0 & 0 & 0 & \dfrac{2.84}{1+0.1s} & 0 & 0 \\
0 & 176.2433 & 0 & 0 & 0 & 0
\end{bmatrix}
$$

Second order plus delay approximation

$$
\begin{bmatrix}
0 & \dfrac{-1.28e^{-0.05s}(1+9.4s)}{(1+2.68s)(1+11.5s)} & \dfrac{-73.4e^{-0.0s}(1+1.1s)}{(1+0.4s)(1+3.7s)} & 0 & \dfrac{-1.82e^{-3.21s}(1+0.0s)}{(1+7.0s)(1+3.1s)} & \dfrac{48.24e^{-0.31s}(1+4.31s)}{(1+2.19s)(1+7.24s)} \\
61.7 & 176.2433 & 0 & 0 & 183.3005 & 0 \\
0 & 176.2433 & 0 & \dfrac{96.7}{1+0.1s} & 183.3005 & 0 \\
0 & 0 & 431.450 & 0 & 0 & 0 \\
0 & 0 & 0 & \dfrac{2.84}{1+0.1s} & 0 & 0 \\
0 & 176.2433 & 0 & 0 & 0 & 0
\end{bmatrix}
$$

Case 12: TBT 105°C and recycle flow 11,500 t/h
First order plus delay approximation

$$
\begin{bmatrix}
0 & \dfrac{-0.982e^{-0.0s}}{(1+2.541s)} & \dfrac{-56.769e^{-0.0s}}{(1+1.96s)} & 0 & \dfrac{-1.3815e^{-5.2904s}}{(1+7.5647s)} & \dfrac{32.09}{(1+4.2s)} \\
61.7 & 176.2433 & 0 & 0 & 183.3005 & 0 \\
0 & 176.2433 & 0 & \dfrac{96.7}{1+0.1s} & 183.3005 & 0 \\
0 & 0 & 431.4492 & 0 & 0 & 0 \\
0 & 0 & 0 & \dfrac{3.0473}{1+0.1s} & 0 & 0 \\
0 & 176.2433 & 0 & 0 & 0 & 0
\end{bmatrix}
$$

Second order plus delay approximation

$$
\begin{bmatrix}
0 & \dfrac{-0.9817(1+2.1666s)}{(1+4.8149s)(1+0.056s)} & \dfrac{-56.769e^{-0.02s}(1+0.64s)}{(1+1.17s)(1+2.44s)} & 0 & \dfrac{-1.3815e^{-3.2258s}(1+0.0318s)}{(1+3.1148s)(1+6.2818s)} & \dfrac{32.09e^{-0.2315s}(1+4.5214s)}{(1+6.6506s)(1+2.058s)} \\
61.7 & 176.2433 & 0 & 0 & 183.3005 & 0 \\
0 & 176.2433 & 0 & \dfrac{96.7}{1+0.1s} & 183.3005 & 0 \\
0 & 0 & 431.4492 & 0 & 0 & 0 \\
0 & 0 & 0 & \dfrac{3.0473}{1+0.1s} & 0 & 0 \\
0 & 176.2433 & 0 & 0 & 0 & 0
\end{bmatrix}
$$

References

Gibilaro, L.G. and F.P. Lees (1969), The reduction of complex transfer function models using the method of memontems, *Chem. Eng. Sci.*, **24**, 85–93.

Jamshidi, M. (1983), *Large Scale Systems: Modeling and Control*, North Holland, New York.

Marshall, S.A. (1966), An approximate method for reducing the order of a linear system, *Int. J. Control*, **10**, 642–643.

Moore, B.C. (1981), Principal component analysis in linear systems: Controllability, observability and model reduction, *IEEE Trans. Automat. Control*, **AC-26**, 17–32.

Safanov, M.G. and R.Y. Chiang (1989), A schur method for balanced-truncation model reduction, *IEEE Trans. Automat. Control*, **34**(7), 729–733.

Wilson, D.A. and R.N. Mishra (1979), Optimal reduction of multivariable systems, *Int. J. Control*, **29**(2), 267–278.

Unbehauen, H. and G.P. Rao (1987), *Identification of Continuous Systems*, North Holland, Amsterdam, Netherlands.

7

Optimal PID Controller Tuning in Large MSF Plants for Seawater Desalination

PID controllers have been widely used in the process industry for more than the past 50 years. The principle of PID controllers seems to have originated in the work of Nicholas Minorsky in connection with the steering of ships in the early 1920s. During the 1940s, Ziegler–Nichols (Z–N) (Ziegler and Nichols, 1942) tuning formulae appeared and have remained in wide use until today in process control. The Z–N tuning rules use some features of the so-called process reaction curve, which is actually the step response curve of the process to be controlled. There are other methods, which use the features of the Nyquist curve. The principle of autotuning is based on the latter, wherein the critical amplitude and frequency at the Nyquist crossover are used in simple tuning rules. Since the early 1980s, commercial autotuners have been designed on the basis of these tuning methods (Smith and Corripio, 1985). The amount of information on the plant dynamics used by these methods being meager, they do not provide good tuning. For example, the heuristic Z–N tuning laws give rise to an oscillatory closed-loop response.

Several approaches to improve PID tuning above the level of quality achieved by the Z–N method have been reported in the literature. In Atherton and Zhunag (1993), a tuning method based on phase margin was reported. In Hang et al. (1991), a refinement of the Z–N method was suggested. In the works of Zhuang and Atherton (1991), the tuning methods are based on optimization in the time domain.

Standard MATLAB routines are used extensively in the computation of the optimal PID controller parameters for the well-known first order with dead-time (FODT) normalized form of plant models. These results are fitted into simple formulae that can be used as ready-made optimal tuning rules. The optimal design is carried out with a wide variety of integral performance criteria such as integral of the squared error (ISE), integral of time-weighted squared error (ISTE), and integral of the absolute value of the error (IAE), integral of time-weighted absolute error (ITAE).

The following are the salient features of the work presented in this chapter:

- A brief review of the PID controller design methods.
- Application of the existing methods to the top brine temperature (TBT) control system in an MSF plant.
- Removal of the limitations existing design methodology, by an efficient simulation-based optimization technique. Removal of the need to approximate the plant model into an FODT form is a significant step.
- The method of optimal PID controllers proposed here uses no approximation of the plant model and is therefore believed to be more accurate than the existing methods.
- By virtue of the experimentally and easily obtainable form in which the plant model information is needed, the proposed method of optimal controller design is of significance to practical applications.
- Multiloop optimization.

7.1 A Brief Review of the Existing PID Controller Tuning Methods

This section describes some methods for determining the parameters of a PID controller. They can be classified broadly into direct and indirect methods. The direct control design techniques are simply prescriptions that tell how the controller parameters should be changed in order to obtain the desired features. The indirect control design methods give controller parameters in terms of model parameters.

7.1.1 Direct Methods

A majority of the PID controllers in the industries are tuned manually by control engineers and operators. The tuning is done based on past experiences and heuristics. By observing the pattern of the closed-loop response to a setpoint change, the operator uses heuristics to directly adjust the controller parameters. The heuristics have been captured in tuning charts that show the responses of the system for different parameter values. When the setpoint changes or major load disturbances occur, properties like damping, overshoot, period of oscillations, and static gains are estimated. Based on these properties, rules for changing the controller parameters to meet desired specifications are executed.

7.1.2 Indirect Methods

7.1.2.1 Ziegler–Nichol's Step Response Method

This method is based on the recording of the open-loop step response of the system, which is characterized by two parameters. This is the unit step response of the controlled plant from which, the information, that is, the values of K_s, θ, and τ, may be obtained and inserted in Table 7.1. This is as shown in Figure 7.1.

The plant open-loop step response is approximated as an FODT form whose transfer function is given as

$$G_{FODT}(s) = \frac{K_s e^{-\theta s}}{(1+\tau s)} \tag{7.1}$$

In using these formulas, we must keep in mind that they are empirical and apply only to limited range of dead time-to-time constant ratios. This means that they should not be extrapolated outside a range of θ/τ to around 1.

TABLE 7.1

Controller Parameters Obtained by the Z–N Step Response Method

Controller	K	T_I	T_D
P	$\dfrac{1}{K_s} \cdot \dfrac{\tau}{\theta}$	—	—
PI	$\dfrac{0.9}{K_s} \cdot \dfrac{\tau}{\theta}$	3.33θ	—
PID	$\dfrac{1.2}{K_s} \cdot \dfrac{\tau}{\theta}$	2θ	0.5θ

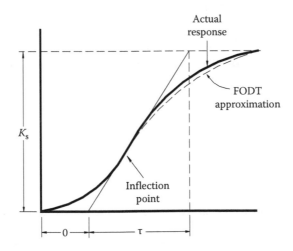

FIGURE 7.1
FODT approximation of the unit step response of the controlled plant.

There are several design methods that are similar to the Ziegler–Nichol's step response method in the sense that they are based on a step response experiment combined with a table that relates the controller parameters to the characteristics of the step response. The most common method is the Cohen–Coon method based on the first order plus dead-time model.

7.1.2.2 Ziegler–Nichol's Frequency Response Method

This method is based on a very simple characterization of the process dynamics. The design is based on knowledge of the point on the Nyquist curve of the process transfer function $G(s)$ where the Nyquist curve intersects the negative real axis. For historical reasons, this point is characterized by the parameters K_{crit} and T_{crit}, which are called the *ultimate gain* and the *ultimate period*, respectively. These two quantities (K_{crit} and T_{crit}) are obtained as follows:

- The control loop is set up with a proportional term alone.
- The controller gain K is raised (or reduced to the proportional band) to bring the loop to the brink of oscillations. The critical value of K at which the closed loop begins to show oscillations is K_{crit}.
- The corresponding period of oscillation is T_{crit}.

The values of K_{crit} and T_{crit} so obtained from the experiment are inserted in Table 7.2 to get the controller parameters K, T_I, and T_D in the respective forms of the controller. If setting up of oscillations on the actual plant is not permitted, the values of K_{crit} and T_{crit} can be indirectly obtained from a computation of the frequency response of the plant model.

7.1.2.3 Optimization Techniques

Optimal PID control can be obtained by minimizing certain integral performance indices when the process model is known. There are various criteria that can be chosen for the purpose of optimization. Some typical ones are

- IAE

$$J_{IAE}(\hat{\theta}) = \int_0^\infty \left| e(\hat{\theta}, t) \right| dt$$

TABLE 7.2

Controller Parameters Obtained by the Z–N Frequency Response Method

Controller	K	T_I	T_D
P	$0.5K_{crit}$	—	—
PI	$0.45K_{crit}$	$0.85T_{crit}$	—
PID	$0.6K_{crit}$	$0.5T_{crit}$	$0.12T_{crit}$

- ISE

$$J_{ISE}(\hat{\theta}) = \int_0^\infty \left[e(\hat{\theta}, t) \right]^2 dt$$

- ISTE

$$J_{ISTE}(\hat{\theta}) = \int_0^\infty t\left[e(\hat{\theta}, t) \right]^2 dt$$

where
the error signal e is the difference between the setpoint and the measurement
$\hat{\theta}$ denotes the variable parameters, which can be chosen to minimize $J_n(\hat{\theta})$ (n = ISE, IAE, ISTE)

7.1.2.4 Tuning PID Controllers Using Critical Gain and Critical Frequency

In obtaining tuning formulae for an FODT plant model, one has to have good estimates of the three parameters, K_s, θ, and τ, for the approximated plant model. Therefore, in this section we determine the formulae for FODT plant models, which enable the optimum ISTE tuning parameters to be found from these measurements of the oscillation frequency and amplitude.

7.1.2.4.1 ISTE Tuning Formulae

To develop these tuning formulae, a relationship must be found between the required PID parameters and critical point data of the plant. The critical point of the Nyquist locus is the point where the phase is 180°, that is, $|G(j\omega_{crit})| = 180°$. The frequency is called the *critical frequency*, and the modulus of the gain is $1/K_{crit}$, where K_{crit} is the critical gain. Thus, for an FODT process, it is easily shown that

$$K_{crit} = \frac{1}{|G(j\omega_{crit})|} \qquad T_{crit} = \frac{2\pi}{\omega_{crit}}$$

from which the values of K_{crit} and ω_{crit} can be calculated. Since for given FODT plant parameters, the plant has a unique critical point and optimal tuning parameters to satisfy the ISTE criteria, it is possible to obtain relationships for the tuning parameters in terms of the critical point parameters

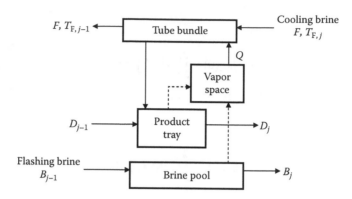

FIGURE 7.2
A typical Nyquist plot.

of the plant. Using a least square fit, the following relations are obtained by Zhuang and Atherton (1991) (Figure 7.2):

$$K = 0.509 K_{crit}$$

$$T_I = 0.412 (3.302 K_s \ K_{crit} + 1) T_D$$

$$T_D = 0.125 T_{crit}$$

Similar relationships are derived using ISTE criterion for PI tuning, and formulae for the PI parameters are

$$K = 0.361 K_{crit}$$

$$T_I = 0.083 (1.935 K_s K_{crit} + 1) T_{crit}.$$

Comparing these formulae with Ziegler–Nichol's frequency response tuning law, it is seen that the proportional gain of the PID controller is reduced from $0.6 K_{crit}$ to $0.509 K_{crit}$, the derivative time is near to the value of the Z–N settings, and the ratio of integral time to the derivative time varies according to the normalized gain ($K_s \ K_{crit}$) instead of remaining fixed. The Ziegler–Nichol's tuning formula was originally derived to give a quarter decay ratio response for load disturbance; therefore, it produces a higher proportional gain, which usually results in a fast relatively high overshoot and oscillatory response for a setpoint change.

Again, the value of proportional gain K is reduced from $0.45 K_{crit}$ to $0.361 K_{crit}$ compared with the Z–N tuning formula, and the integral time varies according to the normalized gain.

The design on the basis of the FODT approximation is plausible, if (θ/τ) lies between 0.1 and 2.0. In this case, there are tables of design information readily available. However, in our present plant, in most cases, the value of (θ/τ) is negligible, rendering the design tables not fully useful.

The FODT model parameters are indeed determined by using the best numerical optimization procedure. Using small values of $\theta/\tau < 0.1$, we will get unduly large values of controller gain, and the related parameters T_I and T_D will also be unsuitable. This is a paradoxical situation, but is understandable, because, amid a lot of optimization and rigor in the model reduction procedure, there is the question for which the answer is not clear:

> How accurately does the FODT approximation preserve the critical point information on the Nyquist curve of the unreduced model?

The Z–N method and the ISTE tuning method are used to determine PI and PID parameters for the MSF process whose critical frequency and gain are calculated and shown in Table 7.3, for the TBT loop for the 12 chosen operating conditions (in this table, K'_c is equal to 120 K_{crit}, where 120 is the maximum span of TBT transmitter).

Table 7.4 lists controller parameters using the (Atherton and Zhunag, 1993) ISTE tuning formulae based on the FODT approximation of $G_{1,6}(s)$.

TABLE 7.3

Critical Points from the Frequency Response of $G_{1,6}$

Case	ω_{crit}	K_{crit} $(1/G(j\omega_{crit}))$	K'_c $120 K_{crit}$	γ $(K_s{}^* K_{crit})$	T_{crit} $(2\pi/\omega_{crit})$	K_s (Plant Steady-State Gain)
1	17.5	2.2090	265.080	119.2843	0.3590	54.00
2	18.2	2.3798	285.576	94.8358	0.3452	39.85
3	21.2	3.6900	442.800	81.3653	0.2964	22.05
4	16.2	2.3321	279.852	74.8368	0.3879	32.09
5	17.9	2.7367	328.404	78.7767	0.3510	28.79
6	19.0	2.9455	353.460	74.6392	0.3307	25.34
7	14.01	1.7443	209.316	106.2453	0.4485	60.91
8	14.6	1.9275	231.300	92.9633	0.4304	48.23
9	14.9	1.9231	230.772	112.7129	0.4217	58.61
10	15.14	1.9417	233.004	88.2891	0.4150	45.47
11	16.1	2.0877	250.524	117.4749	0.3903	56.27
12	16.7	2.3753	289.152	102.2115	0.3762	42.61

TABLE 7.4

Controller Design Based on Z–A Formula for the TBT Loop

Case	TBT (°C)	Rec. Flow (t/h)	FODT Approx. Plant Model $G_{1,6}(s)$	Controller Type	K_P	T_I	T_D
1	95	14,420	$\dfrac{54e^{-0.187s}}{1+5.76s}$	P	—	—	—
				PI	0.7974	6.9074	—
				PID	1.1244	7.3047	0.0449
2	100	14,420	$\dfrac{39e^{-0.32s}}{1+3.79s}$	P	—	—	—
				PI	0.8591	5.2864	—
				PID	1.2113	5.5913	0.0432
3	105	14,420	$\dfrac{22.05e^{-0.391s}}{1+2.3586s}$	P	—	—	—
				PI	1.3321	3.8979	—
				PID	1.8782	5.2330	0.0371
4	105	11,500	$\dfrac{32.09e^{-0s}}{1+4.2s}$	P	—	—	—
				PI	0.8419	4.6944	—
				PID	1.1870	4.9578	0.0485
5	105	12,500	$\dfrac{28.7859e^{-0s}}{1+3.73s}$	P	—	—	—
				PI	0.9880	4.47	—
				PID	1.3930	4.7228	0.0439
6	105	13,500	$\dfrac{25.34e^{-0.028s}}{1+2.697s}$	P	—	—	—
				PI	1.0633	3.9917	—
				PID	1.4993	4.2107	0.0413
7	95	11,500	$\dfrac{60e^{-0.0001s}}{1+6.42s}$	P	—	—	—
				PI	0.6297	7.6899	—
				PID	0.8878	8.1259	0.0561
8	100	11,500	$\dfrac{48e^{-0.0004s}}{1+5.11s}$	P	—	—	—
				PI	0.6958	6.4611	—
				PID	0.9811	6.8255	0.0538
9	95	12,500	$\dfrac{58.6e^{-0.362s}}{1+4.45s}$	P	—	—	—
				PI	0.6942	7.6685	—
				PID	0.9789	8.10433	0.0527
10	100	12,500	$\dfrac{45.5e^{-0.0005s}}{1+4.68s}$	P	—	—	—
				PI	0.7010	5.9191	—
				PID	0.9883	6.2522	0.0519
11	95	13,500	$\dfrac{56.3e^{-0.323s}}{1+4.42s}$	P	—	—	—
				PI	0.7537	7.3954	—
				PID	1.0626	7.8163	0.0488
12	100	13,500	$\dfrac{42.6e^{-0.365s}}{1+3.18s}$	P	—	—	—
				PI	0.8575	6.147	—
				PID	1.2090	6.4949	0.0470

7.2 Proposed Method Using the Unreduced Model Step Response along with Several Integral Performance Criteria

In this section, a simulation facility for PID control loops using step or impulse response of the plant model and methods to tune PID controllers using integral performance criteria is discussed.

In this chapter, the linearized model of an 18-stage MSF desalination plant is considered for six inputs and six outputs. The transfer function matrix of the resulting model is subjected to interaction analysis by the well-known relative gain array (RGA) method and appropriate control structure has been established. The RGA analysis shows the pairings as follows: (u_1, y_2), (u_2, y_6), (u_3, y_4), (u_4, y_5), (u_5, y_3), and (u_6, y_1). The design of an optimal PID controller for one of the most important loops, namely, the TBT loop, is considered in detail.

7.2.1 PID Control Loop Simulation Algorithm with a Nonparametric Process Model

Consider a SISO feedback system with a PID controller as shown in Figure 7.3.

Let the process be described by the sequence $[p_1, p_2, p_3, p_4,...]$ of samples at intervals of $[O, T, 2T, 3T,...]$ of the impulse response function $p(t)$. Define the matrix (Woldai et al., 1995a–c)

$$
\mathbf{P} = T \begin{bmatrix}
\dfrac{p_1}{2} & & & & & & \\
\dfrac{p_2}{2} & \dfrac{p_1}{2} & & & & \mathbf{O} & \\
\dfrac{p_3}{2} & p_2 & \dfrac{p_1}{2} & & & & \\
\dfrac{p_4}{2} & p_3 & p_2 & \dfrac{p_1}{2} & & & \\
\dfrac{p_5}{2} & p_4 & p_3 & p_2 & \dfrac{p_1}{2} & & \\
\cdots & \cdots & \cdots & \cdots & \cdots & \cdots & \cdots \\
\dfrac{p_n}{2} & p_{n-1} & p_{n-2} & p_{n-3} & \cdots & p_2 & \dfrac{p_1}{2}
\end{bmatrix}
$$

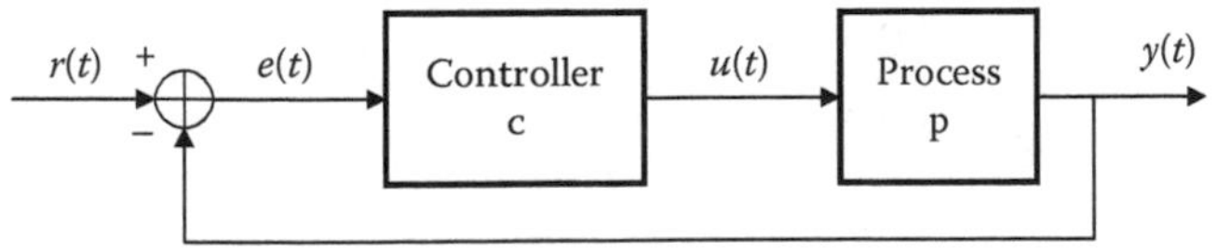

FIGURE 7.3
An SISO feedback control system.

and the vectors of samples of the loop signals

$$\mathbf{r} = \begin{bmatrix} r_1 & r_2 & r_3 & r_4 & \cdots & r_n \end{bmatrix}^T$$
$$\mathbf{e} = \begin{bmatrix} e_1 & e_2 & e_3 & e_4 & \cdots & e_n \end{bmatrix}^T$$
$$\mathbf{u} = \begin{bmatrix} u_1 & u_2 & u_3 & u_4 & \cdots & u_n \end{bmatrix}^T$$
$$\mathbf{y} = \begin{bmatrix} y_1 & y_2 & y_3 & y_4 & \cdots & y_n \end{bmatrix}^T$$

Then,

$$\mathbf{y} = \mathbf{Pu}$$
$$\mathbf{e} = \mathbf{r} - \mathbf{y}$$
$$\mathbf{u} = \mathbf{Ce}$$

where **C** is a matrix like **P** containing the controller information.

7.2.1.1 PID Controller

The output of the controller

$$u(t) = K_C \left[e(t) + \frac{1}{T_I} \int_0^t e(t)dt + T_D \frac{de}{dt} \right]$$

In terms of the operational matrices (Rao, 1983),

$$\mathbf{u} = K_C \left[1 + \frac{T\mathbf{E}}{T_I} + \frac{T_D}{T}\mathbf{D} \right] \mathbf{e}$$

Therefore, for the PID controller

$$\mathbf{C} = K_C \left[1 + \frac{T\mathbf{E}}{T_I} + \frac{T_D}{T}\mathbf{D} \right]$$

Based on trapezoidal rule for integration and backward difference formula for derivative

$$\mathbf{E} = T \begin{bmatrix} 0.5 & & & & & \\ 0.5 & 0.5 & & & \mathbf{O} & \\ 0.5 & 1 & 0.5 & & & \\ 0.5 & 1 & 1 & 0.5 & & \\ \cdots & \cdots & \cdots & \cdots & & \\ 0.5 & 1 & 1 & \cdots & \cdots & 0.5 \end{bmatrix}$$

and

$$D = \begin{bmatrix} -1 & 1 & & & & & \\ 0 & -1 & 1 & & & \mathbf{O} & \\ 0 & 0 & -1 & 1 & & & \\ \dots & \dots & \dots & \dots & \dots & & \\ 0 & 0 & 0 & \dots & \dots & -1 & 1 \end{bmatrix}$$

respectively.

The formulae mentioned give

$$\mathbf{y} = [\mathbf{I} + \mathbf{PC}]^{-1} \mathbf{PCr}$$

where $\mathbf{I}$ is an identity matrix.

The given algorithm can be written in the following recursive form for implementation in the desired simulation routine:

$$y_1 = \frac{T}{2} p_1 K_C \alpha r_1 / \beta$$

where

$$\alpha = \left(1 + \frac{T}{2T_I} + \frac{T_D}{T}\right); \quad \beta = \left(1 + \frac{T}{2} p_1 K_C \alpha\right)$$
$$e_1 = r_1 - y_1$$
$$u_1 = K_C \alpha e_1$$

Remark: While developing a recursive formula, direct expansion of the matrix version is avoided, because *D* is the form that brings in a future value (at $k = 2$) in the starting expression corresponding to $k = 1$. This is against the principle of recursion. To circumvent this problem, one should consider only the backward differences for the derivative. At $k = 1$, this implies the use of values at $k = 0$, which are all zeros and fall out of the defined vectors of signals. Thus, delete the first column of *D* and expand the matrix formula. This operation leaves no effect on the result, since all the related values of signals at $k = 0$ are zeros.

$$y_2 = \frac{\left(\frac{T}{2} p_2 u_1 + \frac{T}{2} p_1 K_C \left[\frac{T}{2T_I} - \frac{T_D}{T}\right] e_1 + \frac{T}{2} p_1 K_C \alpha r_2\right)}{\beta}$$

$$e_2 = r_2 - y_2$$

$$u_2 = K_C \left[e_2 \frac{T}{2T_I} + (e_1 + e_2) + \frac{T_D}{2T}(e_2 - e_1) \right]$$

and for $k = 3, 4, \ldots$

$$y_k = \frac{\left\{ \frac{T}{2} p_k u_1 + T \sum_{j=2}^{k-1} p_{k-j+1} u_j + \frac{T}{2} p_1 K_C \left[\frac{T}{2T_I} e_1 - \frac{T_D}{T} e_{k-1} + \frac{T}{T_I} \sum_{k=3}^{n} e_{k-1} \right] + \frac{T}{2} p_1 K_C \alpha r_k \right\}}{\beta}$$

$$e_k = r_k - u_k$$

$$u_k = K_C \left[e_k + \frac{T}{2T_I}(e_1 + e_k) + \frac{T}{T_I} \sum_{3}^{n} e_{k-1} + \frac{T_D}{T}(e_k - e_{k-1}) \right]$$

If the plant model data are available in the form of a sequence of values of the step response function $\{s_i, \; i = 1,2,\ldots\}$, the sequence of impulse response values required for the proposed algorithm can be computed as

$$\left\{ p_k = \frac{s_{k+1} - s_k}{T}, \quad k = 1,2,\ldots \right\}$$

Alternatively, we may replace the sequence $\{p_i\}$ with $\{s_i\}$ and drive the plant with the sequence $\{\Delta u_k\}$ where $\Delta u_k = \{u_k - u_{k-1}\}/T$. With either of these afore-mentioned modifications, the algorithm can work with the sequence of samples of the step response function.

7.2.2 Optimal PID Controller Design for TBT Control in an 18-Stage MSF Desalination Plant

The nonlinear plant model is linearized at an operating point choosing six manipulated variables (input) and six controlled variables (outputs) into a standard state-space form. The linearized model has 155 state variables, which was processed through MATLAB and the impulse response function was evaluated (Woldai et al., 1996) (Figure 7.4).

Figure 2.5 shows the brine heater and steam flow schemes in the plant, which are the most important control loops in the MSF desalination process that directly affect the production. We consider the brine heater loop controlling the brine outlet temperature denoted as TBT. This is controlled by manipulating the steam flow into the brine heater by means of a steam valve in this setup $u(t)$. The steam value position (controller output) is the manipulated variable and $y(t)$ is the TBT.

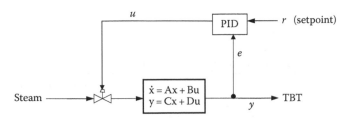

FIGURE 7.4
Closed-loop TBT controller system.

Use of an integral performance criterion often results in a better closed-loop response of a control system than heuristic tuning methods, since the method takes into account the whole transient response of the system. The approaches are based on optimization with suitable criteria, which will be seen to be appropriate in such energy-intensive controls.

Table 7.5 provides the optimal PID controller parameters for plant operation at the chosen point. Figure 7.5 presents the corresponding unit step response of the optimally tuned TBT control loop, and similarly, Figure 7.6 gives the corresponding unit step response function of the optimally controlled closed-loop system, in the operating region of the six cases. The Z–A method is not included within this set of results as this method may not applicable in this case. The ISTE, ISE, and IAE criterion methods give better closed-loop responses with a smaller overshoot.

A simple and reliable robust algorithm of simulation of feedback control system with PID controllers and process models in the form of unit impulse or step response functions is presented. The PID controllers obtained by simulation using this algorithm are truly optimal and compared (Ogata, 2011) with those based on FODT approximations. The tuning algorithms (Lopez et al., 1967; Corripio, 1990) have been tried out on several different cases (Woldai et al., 1995a–c) or plant models, in all cases have produced better results.

7.3 Multiloop Optimization

The method presented in this book is applicable to optimal design of multivariable processes in which we can optimize the whole set of control loops in a process if such a procedure is warranted. Our discussion on optimal control of MSF processes has so far been based on single-loop design optimization. The study took up the design of the isolated TBT control loop in view of the fact that the RGA analysis of the model of the plant has indicated no need for multivariable design.

TABLE 7.5

Optimal PID Settings Based on an Unreduced Plant Model

Case	TBT (°C)	Rec. Flow (t/h)	Criterion	K_P	T_I	T_D
				Parameters of PID Controllers		
1	95	14,420	ISE	0.7346	2.6647	0.2734
			IAE	0.8007	5.3364	0.1814
			ISTE	0.8246	6.5527	0.0809
2	100	14,420	ISE	0.8120	2.5166	0.2521
			IAE	0.9087	3.9736	0.1541
			ISTE	0.8654	4.2280	0.1727
3	105	14,420	ISE	1.2292	2.05	0.2219
			IAE	1.2162	3.0922	0.1226
			ISTE	1.2893	3.0901	0.1506
4	105	11,500	ISE	0.8108	1.8366	0.3492
			IAE	0.8271	2.8149	0.2307
			ISTE	0.8678	2.8599	0.2304
5	105	12,500	ISE	0.8547	1.8909	0.2921
			IAE	0.9050	2.7296	0.1918
			ISTE	0.9002	2.7488	0.2038
6	105	13,500	ISE	1.1174	1.8207	0.2874
			IAE	1.1301	3.0086	0.1632
			ISTE	1.1906	3.0579	0.1844
7	95	11,500	ISE	0.5313	2.8573	0.3946
			IAE	0.4867	4.1303	0.2758
			ISTE	0.5741	4.3894	0.2807
8	100	11,500	ISE	0.5756	2.4412	0.3745
			IAE	0.6035	3.5802	0.2219
			ISTE	0.6144	3.5914	0.2670
9	95	12,500	ISE	0.6504	3.0501	0.4104
			IAE	0.7034	4.8765	0.2570
			ISTE	0.6938	4.8380	0.2812
10	100	12,500	ISE	0.5957	2.4129	0.3427
			IAE	0.5449	3.4296	0.2370
			ISTE	0.6298	3.4345	0.2502
11	95	13,500	ISE	0.6773	2.8379	0.3760
			IAE	0.7172	4.6499	0.2311
			ISTE	0.7273	4.6982	0.2573
12	100	13,500	ISE	0.7892	2.4683	0.3720
			IAE	0.7897	3.7862	0.2411
			ISTE	0.8363	3.7372	0.2502

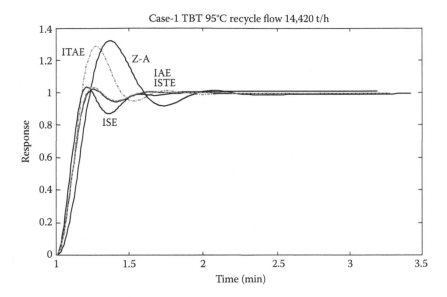

FIGURE 7.5
(**See color insert.**) Step response of the optimally tuned TBT control loop.

The procedure for optimization can be easily extended to the multivariable situation shown in Figure 7.7 if the interactions among the loops warrant such a treatment (Seborg et al., 1989). This may be done as follows:

1. Apply the RGA analysis and choose the *recommended* input–output pairing.
2. Define an appropriate optimality criterion including the errors in all the loops. If J_k is the performance index for the kth loop in terms of its error e_k, then the overall performance index may be taken as the sum of the indices of all the loops, implying uniform weighting for all the loops.

$$J = \sum_{k=1}^{\text{No. of loops}} J_k$$

3. Depending upon the relative importance of the individual loops, different weights may also be assigned.
4. Optimize J with respect to the parameters of the controllers of all the loops simultaneously.

Naturally, the multivariable situation requires simulation of all the loops. The computational burden of the optimization procedure also increases with the number of loops.

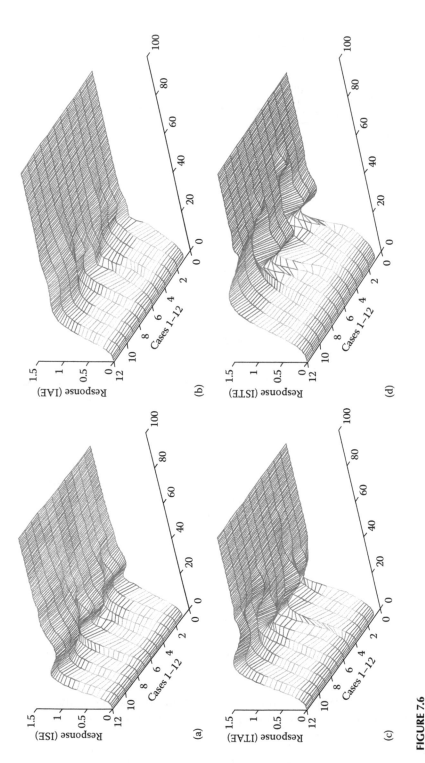

FIGURE 7.6
(See color insert.) Step response of the optimally tuned systems in the operating space: (a) responses for optimal PID settings using ISE criterion, (b) responses for optimal PID settings using IAE criterion, (c) responses for optimal PID settings using ITAE criterion, and (d) responses for optimal PID settings using ISTE criterion.

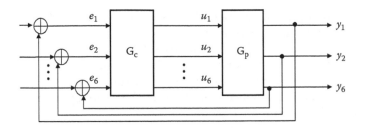

FIGURE 7.7
Multi-input-multi-output (MIMO) system.

TABLE 7.6

Controller Settings

Controller Pairing	Tuning Method	K_C	T_c	T_D
$u_6 \rightarrow y_1$	Single loop/ISE	0.7346	2.6647	0.2734
$u_4 \rightarrow y_5$	Single loop/ISE	0.5	0.65E−8	0.017
$u_6 \rightarrow y_1$	Multiloop/ISE	0.7346	2.6639	0.2733
$u_4 \rightarrow y_5$	Multiloop/ISE	1.385	0.12267E−7	0.030428

The procedure was applied to the case of the present MSF plant with two inputs and two outputs based on the ISE criterion. The input–output pairing is made as suggested by the RGA method. Although no interactions are indicated, the multivariable optimization procedure with uniform weighting of errors in the two loops has been applied. The optimal PID controller parameters are shown in Table 7.6.

The results in this case, which is one of decoupled loops, are as expected—the same as those obtained by single-loop optimization. This demonstrates and verifies the validity of the optimization procedure suggested in this book.

Nomenclature

M_B	Flashing brine holdup in a stage
B_i	Brine inlet flowrate to a stage
B_o	Brine outlet flowrate from a stage
$C_{B,o}$	Salt mass fraction in the brine leaving a stage
$C_{B,i}$	Salt mass fraction in the brine entering a stage
$V_{B,o}$	Vapor generation rate from flashing brine in a stage
$H_{B,o}$	Enthalpy of flashing brine leaving a stage
$H_{B,i}$	Enthalpy of flashing brine entering a stage
H_{VB}	Enthalpy of vapors leaving the flashing brine in a stage

$T_{B\text{-sat}}$	Saturation temperature of flashing brine leaving a stage
T_B	Flashing brine temperature leaving a stage
T_{NEA}	Nonequilibrium allowance temperature
$P_{B\text{-sat}}$	Saturation pressure of flashing brine in a stage
P_B	Total pressure over flashing brine in a stage
Y_{IM}	Mole fraction of noncondensables in the vapor space of a stage
M_D	Distillate holdup in a stage
D_i	Distillate inlet flowrate to a stage
D_o	Distillate outlet flowrate from a stage
F_C	Total condensate flow, falling into distillate channel in a stage
$V_{D,o}$	Vapor generation rate from distillate tray in a stage
$H_{D,i}$	Enthalpy of distillate entering a stage
$H_{D,o}$	Enthalpy of distillate leaving a stage
H_{VD}	Enthalpy of vapors leaving the flashing brine in a stage
$T_{D,o}$	Temperature of distillate leaving a stage
T_{V_sat}	Saturation temperature in the vapor space of a stage
M_V	Vapor and noncondensable holdup in a stage
F_I	Flowrate of inerts (carbon dioxide and air)
V_i	Vapor entering a stage from previous stage
V_o	Vapor leaving a stage (next stage or vent)
$Y_{I,i}$	Mass fraction of noncondensables in the vapor space of a stage
$P_{V\text{-sat}}$	Saturation pressure in the vapor space
P_V	Total pressure in the vapor space
K_{DEM}	Demister pressure drop constant
V_{Bi}	Vapor velocity in the demister of a stage
$H_{VV,i}$	Enthalpy of vapor entering from previous stage
$H_{VV,o}$	Enthalpy of vapor going toward the cooling tubes in a stage
H_{VD}	Enthalpy of vapors leaving the distillate tray
F_i	Cooling brine inlet flowrate to a stage
F_o	Cooling brine outlet flowrate from a stage
M_F	Cooling brine holdup in a stage
$H_{F,i}$	Enthalpy of cooling brine entering a stage
$H_{F,o}$	Enthalpy of cooling brine leaving a stage
Q	Heat transferred to cooling brine in a stage
U	Overall heat transfer coefficient in a stage
T_{im}	Log mean temperature in a stage
A_t	Area (surface of tubes)
$T_{F,i}$	Temperature of cooling brine entering a stage
$T_{F,o}$	Temperature of cooling brine leaving a stage
R_e	Recycle brine flowrate
C_{sea}	Salt mass fraction in the seawater
H_M	Enthalpy of makeup
B_D	Brine blowdown flowrate
M	Makeup flowrate
A_s	Stage bottom area

L_B	Flashing brine level in a stage
ρ_B	Density of brine in a stage
A_D	Distillate channel bottom area in a stage
L_D	Distillate level in a stage
ρ_D	Density of distillate in a stage
V	Vapor space volume in a stage
ρ_V	Density of vapors in the vapor space
F_S	Steam flowrate to brine heater
F_{CS}	Total condensate flow from brine heater
$H_{S,i}$	Enthalpy of steam entering the brine heater
$H_{C,o}$	Enthalpy of condensate in the brine heater
Q_H	Heat transferred to heating brine in the brine heater
$H_{t,o}$	Enthalpy of cooling brine leaving the brine heater
$H_{t,i}$	Enthalpy of cooling brine entering the brine heater
A_H	Total heat transfer area in the brine heater based on the tube outside diameter
T_S	Temperature of steam entering the brine heater
T_{ave}	Average temperature
P_1	Vapor pressure before orifice gate
P_2	Vapor pressure after orifice gate
H_o	Orifice height
F_{max}	Flow at full opening
R_{eff}	Effective rangeability
K_s	Steady-state gain
θ	Dead time
τ	Time constant ratios
K	Controller gain
T_I	Integral time
T_D	Derivative time

References

Astrom, K.J., T. Hagglund, C.C. Hang, and W.K. Ho (1993), Automatic tuning and adaptation for PID controllers—A survey, *Control Eng. Pract.*, 1(4), 699–714.

Åstrom, K.J. and B. Wittenmark (1989), *Adaptive Control*, Addison-Wesley, Reading, MA.

Atherton, D.P. and M. Zhuang (May 1993), Automatic tuning of optimum PID controllers, *IEE Proc. D*, 140(3), 216–224.

Corripio, A.B. (1990), *Tuning of Industrial Control Systems*, ISA, Raleigh, NC.

Hang, C.C., K.J. Astrom, and W.K. Ho (1991), Refinements of Ziegler–Nichol's tuning formula, *IEE Proc. D*, 138(2), 111–118.

Lopez, A.M., P.W. Murril, and C.L. Smith (November 1967), Controller tuning relationships based on integral performance criteria, *Instrum. Technol.*, 14(11), 57.

Ogata, K. (2011), *Modern Control Engineering*, 5th edn., Eastern Economy Edition, Prentice Hall, India, pp. 601–603.

Rao, G.P. (1983), *Piecewise Constant Orthogonal Functions and their Application to Systems and Control*, Lectures notes in Control and Information Sciences, Springer-Verlag, Berlin, Germany.

Seborg, D.E., D.A. Mellichamp, T.F. Edgar, and F.J. Doyle, III (2010), *Process Dynamics and Control, 3rd Edition*, John Wiley and Sons, Hoboken, NJ.

Smith, A.C. and A.B. Corripio (1985), *Principles and Practice of Automatic Process Control*, John Wiley & Sons, New York.

Tsypkin, Y.Z. (1978), Algorithms of optimization with a priori uncertainty (past, present, future), in *Proceedings of the Sixth IFAC Congress*, Helsinki, Finland, Pergamon Press, Oxford, UK.

Woldai, A., D.M.K. Al-Gobaisi, R.W. Dunn, A. Kurdali, and G.P. Rao (1995a), Simulation aided design and development of an adaptive scheme with optimally tuned PID controller for large MSF seawater desalination plant (part I), in *Proceedings of the Fifth IFAC Symposium on Adaptive System in Control and Signal Processing*, Budapest, Hungary, June 14–16, 1995.

Woldai, A., D.M.K. Al-Gobaisi, R.W. Dunn, A. Kurdali, and G.P. Rao (1995b), Simulation aided design and development of an adaptive scheme with optimally tuned PID controller for large MSF seawater desalination plant (part II), in *Proceedings of the Fifth IFAC Symposium on Adaptive System in Control and Signal Processing*, Budapest, Hungary, June 14–16, 1995.

Woldai, A., D.M.K. Al-Gobaisi, R.W. Dunn, A. Kurdali, and G.P. Rao (1995c), Simulation aided design and development of an adaptive scheme with optimally tuned PID controller for large MSF seawater desalination plant (part III), in *Proceedings of the Fifth IFAC Symposium on Adaptive System in Control and Signal Processing*, Budapest, Hungary, June 14–16, 1995.

Zhuang, M. and D.P. Atherton (1991), Tuning PID controllers with integral performance criteria, *Proc. IEE Conf. Control*, **1**(332), 481–486.

Ziegler, J.G. and N.B. Nichols (1942), Optimum settings for automatic controllers, *Trans. ASME* **64**, 759–768.

8

An Adaptive Scheme with an Optimally Tuned PID Controller

Some of the basic problems in process industries are the variations in the plant characteristics, which occur due to changing operating conditions and unforeseen disturbances acting on the plant. An adaptive control system is one in which the controller parameters are adjusted automatically in such a way as to compensate for such variations. Studies by simulation on the model of an MSF desalination plant have shown certain features of the model developed on the basis of physical laws and correlations. The most important of these are related to nonlinearity, which results in changing of process characteristics. This study showed a variation in the parameters of the linearized model with the operating conditions. It is clear that a fixed PID controller cannot be optimal if the operating point changes from the one at which the optimal controller was designed. Therefore, adaptive control schemes are used, where the controller parameters are updated to match the plant parameters. In this chapter, two adaptive control schemes are presented. One is based on a linear parameter scheduling law and the other employs an artificial neural network (ANN) to tune the controller. Appendix 8.A provides a brief introduction to the ANN of interest here.

8.1 Adaptive Tuning Techniques

There are many different approaches to adaptive control, and in fact, there are many different definitions of adaptive control. In addition, there are several ways of classifying adaptive control approaches. One useful distinction is between direct adaptive controllers, in which control law parameters are determined directly from the parameters that are determined in terms of the identified mode. Since other texts can give a more extensive review of the range of adaptive techniques (Åstrom and Wittenmark, 1989; Tsypkin, 1978), the intent here is to mention those which are most widely used presently in related processes. The field of adaptive control has received much attention during the last few years because of the increasing availability of relatively inexpensive computing power. Various adaptive techniques have been implemented in single-loop PID controllers during the last decade.

The various types of adaptive control systems (Åstrom et al., 1993) differ only in the way the parameters of the controller are adjusted. It is best to choose a method by considering the actual conditions in a practical situation, which warrant a change in the settings of the controller. For practical applicability, simplicity as much as possible is very essential. In the case of the MSF plant, the study hitherto included physical modeling and controller design by optimization. As there was no provision for parameter estimation, self-tuning control type methods could not be applied in the hopes of achieving adaptation under the general conditions of uncertainty although we did discuss an important means of reducing uncertainties in the process measurements by data reconciliation.

An examination of the nature of the model revealed the nonlinear character of the process. It is appropriate to advance the controller to suit this character in a simple way. Since designing controllers for nonlinear plants is quite complex, we took the path of adaptation by parameter scheduling to meet the needs of the problem. In this chapter, an adaptive control scheme based on the linear parameter scheduling law and ANN is discussed.

8.2 Parameter Scheduling

Parameter scheduling is shown to be a useful alternative to continuously adaptive techniques where robustness is required with changing plant conditions; for parameter scheduling to be successful, the changes in process parameters must be predictable or at least a repeatable, function of measurable process parameters. The objective is to adjust controller parameters in response to changing process dynamics or disturbance dynamics. In parameter scheduling, an auxiliary variable is used to find the best values of controller parameters as shown in Figure 8.1.

If there is an auxiliary variable that correlates well with the changes in the process dynamics, it can be related ahead of time to the *best* values of the controller parameters.

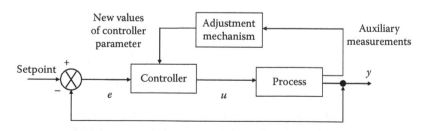

FIGURE 8.1
Parameter scheduling concept.

The methods suggested in this work are close in spirit to parameter scheduling. Referring to Figure 8.1, the current values of TBT and the recycle flow, namely, T and F, play the role of *auxiliary measurements*. The two methods suggested here differ in the way the *adjustment mechanism* is realized in the process *parameter scheduling*. The basis of parameter scheduling in both is a set of optimal design conditions corresponding to 12 operating conditions (Cases 1–12) obtained by simulation as described in the previous chapters. It should be stated that the effectiveness of these will heavily depend on the quality of the plant model, which in the present case is assured, by virtue of validation by the measurements on the actual plant.

8.2.1 The Space of Operating Conditions of the Plant

Based on the studies and plant operating experience, the most important of the plant operating conditions to be enlisted in the space are as follows:

T is the top brine temperature (TBT)

F is the brine recycle flowrate

These two are fixed according to the requirements of the plant production rate simultaneously satisfying other important conditions such as performance ratio. There are limits set on these variables for practical reasons. For example, an upper limit on T is set in view of plant vulnerability for scaling and a lower limit on the velocity of brine through tubes (thereby on F) to avoid sludge formation. Likewise, a lower limit on T and a higher limit on the velocity of brine in the tubes (thereby on F) are set based on certain other conditions. If we denote the ranges of TBT and brine recycle flow by $[T_{min}, T_{max}]$ and $[F_{min}, F_{max}]$, respectively, and consider a set of operating conditions $\{T_k, F_k, k = 1, 2, ..., N\}$ in this region, we obtain the optimal PID controller parameter vector $\mathbf{c}_k = [K_{pk} T_{ik} T_{dk}]^T$ for $k = 1, 2, ..., N$ by the simulation facility developed earlier. Twelve points have been generated in this space. They relate the operating conditions with the optimal controller settings as shown in Figure 8.2.

8.2.2 Approximate Mapping of the Plant Operating Condition Space into the Controller Parameter Space

The 2D space of plant operation conditions is sampled as

$$\{T_k; F_k, \quad k = 1, 2, ..., N\}$$

At each of these points, the controller parameter vectors are given by

$$\mathbf{c}_k = \left[k_{pk} T_{ik} T_{dk} \right]^T$$

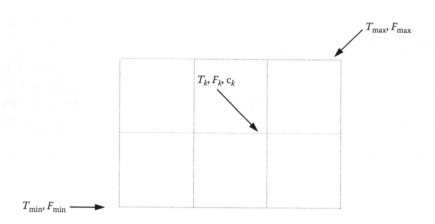

FIGURE 8.2
Space of operating conditions.

In view of the not too wide variations in T and F, we consider a mapping function of the form

$$\mathbf{c} = \mathbf{f}(T,F)$$

or

$$c_1 = f_1(T,F) = a_{11} + a_{12}T + a_{13}F + a_{14}TF = k_\mathrm{p}$$

$$c_2 = f_2(T,F) = a_{21} + a_{22}T + a_{23}F + a_{24}TF = T_\mathrm{i}$$

$$c_3 = f_3(T,F) = a_{31} + a_{32}T + a_{33}F + a_{34}TF = T_\mathrm{d}$$

Using the relations established at the points 1, 2, ..., N, the coefficients in the vector function $\mathbf{f}$ are determined by least squares fitting to form the parameter scheduling law for a chosen optimization criterion.

A convenient form of this adaptive law is

$$\mathbf{c} = \begin{bmatrix} a_{11} & a_{12} & a_{13} & a_{14} \\ a_{21} & a_{22} & a_{23} & a_{24} \\ a_{31} & a_{32} & a_{33} & a_{34} \end{bmatrix} \begin{bmatrix} 1 \\ T \\ F \\ TF \end{bmatrix} = \begin{bmatrix} k_\mathrm{p} \\ T_\mathrm{i} \\ T_\mathrm{d} \end{bmatrix}$$

8.2.3 Parameter Scheduling Law for an 18-Stage MSF Desalination Plant

Table 7.5 provides the PID controller parameters obtained by simulation on the unreduced plant. In the parameter scheduling strategies obtained by least squares fitting (Woldai et al., 1996), the results in the case of the four optimizing criteria are given by the following:

$$ISE: \quad c = \begin{bmatrix} 7.574 & -0.0815 & -0.0008 & 0.0000 \\ 29.695 & -0.2732 & -0.0014 & 0.0000 \\ 1.6834 & -0.0086 & 0.0000 & 0.0000 \end{bmatrix} \begin{bmatrix} 1 \\ T \\ F \\ TF \end{bmatrix}$$

$$IAE: \quad c = \begin{bmatrix} 1.6087 & -0.02732 & -0.003 & 0.0000 \\ -10.6906 & 0.1181 & 0.0025 & 0.0000 \\ 0.5822 & 0.0001 & 0.0000 & 0.0000 \end{bmatrix} \begin{bmatrix} 1 \\ T \\ F \\ TF \end{bmatrix}$$

$$ISTE: \quad c = \begin{bmatrix} 6.0556 & -0.0668 & -0.0007 & 0.0000 \\ -10.5820 & 0.1155 & 0.0026 & 0.0000 \\ 5.7982 & -0.0507 & -0.0004 & 0.0000 \end{bmatrix} \begin{bmatrix} 1 \\ T \\ F \\ TF \end{bmatrix}$$

8.3 Application of Artificial Neural Networks

8.3.1 ANN-Based Control Loop Optimization

Here, an ANN-based method is proposed to automate the optimization procedure through interpolation. In this approach, it starts with the optimized controller settings at a number of key points in the operating region. These controller settings are obtained by actual optimization through simulation as described in Chapter 7. Twelve key points (Cases 1–12) in the operating region have been chosen. The (T, F) pairs and the corresponding optimal PID controller parameter values have been used to train an ANN by backpropagation learning. More points may be used if necessary to improve learning. Figure 8.3 shows the situation for ANN training. The ANN has been trained with the 12-point information separately for each optimal criterion J_0.

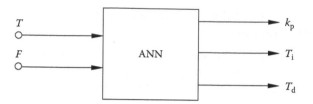

FIGURE 8.3
Generation of training data in the (T, F) operating region.

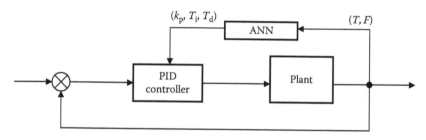

FIGURE 8.4
Automated controller optimized at every operating point by means of an ANN.

The ANN, so trained by the results of optimization at the key points in the operating region, is inserted in the system as shown in the scheme in Figure 8.4. For any point (T, F) in the operating region, the set (k_p, T_i, T_d) is delivered automatically by the ANN for setting the controller parameters optimally at the point.

8.3.2 Comparison of the Methods

Figure 8.5 shows the performance of the two methods proposed here. The 12 points in the 2D space of operating conditions are shown along the horizontal axis of each of the four parts of this figure. This single-dimensional representation of the 2D information in the horizontal axis is the reason behind the apparent discontinuities of the actually smooth adaptation schedule functions. A 3D representation in the space of T, F, c may be made to show the surface of the parameter scheduling law. But this was not done, and the present unconventional representation was made only in the interest of simplicity and visual clarity of the effectiveness of the proposed technique. The mapping and the ANN are both able to model the parameter scheduling law without much difference. This is due to the close sampling of the operating region.

This mapping involves a simple formula representing the parametric scheduling law, which is adequate if the actual optimal control surface is nearly linear. On the other hand, the ANN has the ability to model a nonlinear control

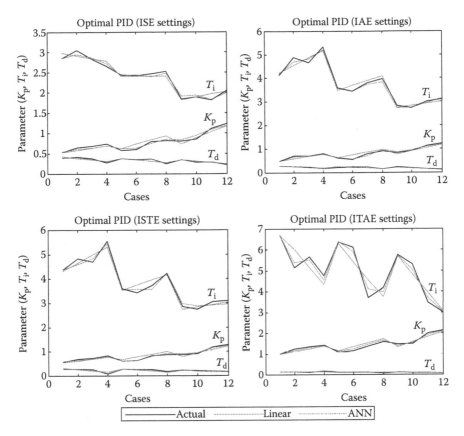

FIGURE 8.5
(See color insert.) Actual and predicted outputs.

surface, that is, the relation between the auxiliary measurements and optimal controller parameters. A block diagram of the overall integrated scheme with parameter scheduling law is shown in Figure 8.6.

The control strategy proposed here is to handle the nonlinear character of the plant. The present parameter scheduling law is expected to be of use in conjunction with a steady-state optimization program at a higher level of hierarchy to provide optimal operating conditions as setpoint values for T and F. These will automatically schedule the controller parameters according to the laws presented here to ensure dynamic optimization according to the chosen integral performance criterion. A potential advantage of the ANN technique of adaptation is the possibility for online training for adaptation on the basis of any other underlying controller design. As long as gaining insight into the performance of the system is not an essential issue and maintaining good performance is all that is desired, the ANN technique is more attractive for its simplicity despite the difficulty in the mathematical analysis of the resulting overall adaptive system. An ANN can also be

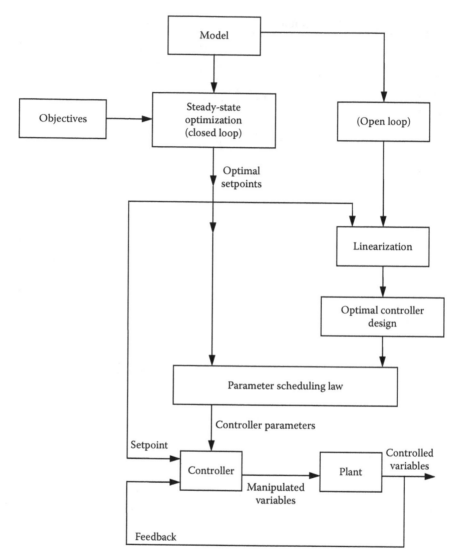

FIGURE 8.6
Integrated adaptive control scheme.

trained to mimic an expert human operator after a course of online training with the system operated by an expert operator.

The mapping method and the trained ANN method are based conceptually on the same principle, and the difference between them is only in the means of realization. A trained ANN is an effective means of capturing nonlinear relations whose formal mathematical descriptions in functional form become unwieldy and often intractable.

8.A Appendix: An Introductory Note on Neural Networks

8.A.1 Artificial Neural Networks

ANNs consist of highly interconnected simple processing elements called *neurons*. A block diagram of a typical neuron is shown in Figure 8.A.1.

Each neuron consists of a summing junction, which adds together the weighted inputs from the other neurons, and an activation function, which generates the neuron output from the summing junction output. The output from the neuron is directed as inputs to other neurons. Neurons transmit signals to each other via weighted links, which attenuate or amplify the transmitted signal depending on the value of the weight.

ANNs can be grouped into various classes depending on their feedback link connection structure or various architectures. In this work, we use the feedforward neural network (FNN) with backpropagation of training. An FNN consists of layers of neurons with weighted links connecting the outputs of neurons in one layer to the inputs of neurons in the next layer. One or more layers exist between the input layer and the output layer. These layers are called the *hidden layers*. A typical feedforward network is shown in Figure 8.A.2. A three-layered network is shown, but in principle, there could be more than one hidden layer.

The output of units in layer 1 is multiplied by appropriate weights w_{ij}, and these are fed as inputs to the next layer, the hidden layer. If i is the input layer, then the output of a unit O_i will be equal to the input of that particular unit, that is, $O_1 = X_1$. The total input to a unit in layer j is

$$Net_j = \sum_i w_{ij}O_i \qquad (8.A.1)$$

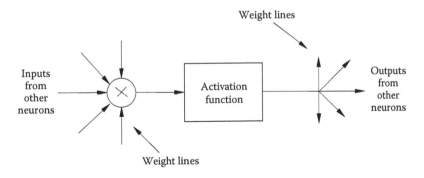

FIGURE 8.A.1
Block diagram of a neuron.

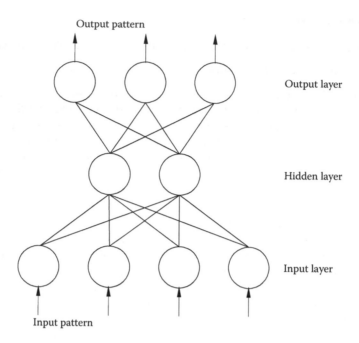

Output pattern

Output layer

Hidden layer

Input layer

Input pattern

FIGURE 8.A.2
Multilayer FNN.

and the output of a unit in layer *j* is

$$O_j = f(Net_j) \tag{8.A.2}$$

where *f* is an activation function. A convenient logistic activation function is given by the relation

$$V_j = f(Net_j) = \frac{1}{\left[1 - \exp - Net_j - \left(\Phi_j / \Phi_0\right)\right]} \tag{8.A.3}$$

The function *f* yields an output that varies continuously from 0 to 1. The quantity Φ_j serves as a *threshold* and positions the transition region of the *f* function. The quantity Φ_0 denotes the abruptness of this transition. An example of such a function is shown in Figure 8.A.3.

In learning the hidden representation (i.e., the weights and the threshold values), the network functions solely in a feedforward manner as shown in Equations 8.A.1 and 8.A.2. In the learning process, the network is fed with two sets of patterns: an input pattern and a corresponding output pattern. Using the (initialized) weights and threshold values, the network produces

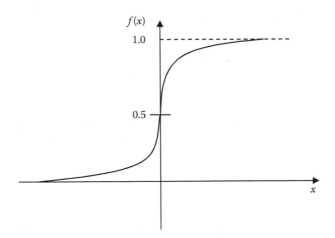

FIGURE 8.A.3
Sample sigmoidal function.

its own output pattern that is then compared with the desired (target) output pattern. The error at any output unit in layer k is

$$e_k = t_k - O_k \tag{8.A.4}$$

where
  t_k is the desired output for that unit in layer k
  O_k is the actual output

The total error function may be written as

$$E = \frac{1}{2} \sum (t_k - O_k)^2 \tag{8.A.5}$$

Learning comprises changing the weights and thresholds so as to minimize the error function in a gradient descent manner. The analytical continuous nature of the activation function allows errors to be traced backward. For the activation function of Equation 8.A.3, the *delta rule* of iterative convergence toward improved values for the weights and the threshold may be stated as

$$\Delta w_{kj} = \eta d_k O_j \tag{8.A.6}$$

where the error signal d_k at an output unit k is given by

$$d_k = (t_k - O_k) O_k (1 - O_k) \tag{8.A.7}$$

and the error signal d_j for an arbitrarily hidden u_j is given by

$$d_j = O_j(1 - O_j)\sum_k d_k w_{kj} \qquad (8.A.8)$$

In Equation A8.6, η is called the *learning rate parameter*. In practice, it has been found that one way to increase the learning rate without causing oscillations is to modify Equation A8.6 to include a momentum term α, that is,

$$\Delta w_{kj}(n+1) = \eta d_k O_j + \alpha \Delta w_{kj}(n) \qquad (8.A.9)$$

where
  n is the number of times for which a set of input patterns has been presented to the network
  α is a constant, which relates how the past weights change to the present ones.

References

Åstrom, K.J., T. Hagglund, C.C. Hang, and W.K. Ho (1993), Automatic tuning and adaptation for PID controllers—A survey, *IFAC J. Control Eng. Pract.*, **1**(4), 699–714.

Åstrom, K.J. and B. Wittenmark (1989), *Adaptive Control*, Addison-Wesley, Reading, MA.

Tsypkin, Y.Z. (1978), Algorithms of optimization with a priori uncertainty (past, present, future), in *Proceedings of the Sixth IFAC Congress*, Helsinki, Finland. Pergamon Press, Oxford, UK.

Woldai, A., D.M.K. Al-Gobaisi, R.W. Dunn, A. Kurdali, and G.P. Rao (1996), An adaptive scheme with an optimally tuned PID controller for a large MSF desalination plant, *Control Eng. Pract.*, **4**(5), 721–734.

9

Conclusions and Discussion

9.1 Conclusions

This book is focused on modeling, simulation, and control of large-scale multi-stage flash (MSF) desalination processes, which produce freshwater from seawater. It includes mathematical modeling, simulation, and prospects of advanced control for the MSF plants, supported by test data from an 18-stage production unit. A system theoretic analysis of a representative model obtained from these studies is the basis of further investigations on appropriate control strategies.

Present-day MSF plant control is accomplished by simple PID schemes, which have remained in rudimentary form for many decades. This book looks at possible improvements on the existing scenario in terms of certain advances in optimal tuning control, adaptive control, as may be warranted by the behavioral complexities of the plant dynamics.

Following a brief introduction to and a discussion of the salient features of the operation and control of MSF plants, the development of a dynamic model based on physical principles has been presented. An 18-stage MSF plant, operating in the UAE, has been considered for the entire study. The model has been validated by actual measurements on this plant. The model so developed has been used in extensive simulations, which showed nonlinear behavior. When linearized at different operating points in the operating region, significant variations in the linearized model parameters have been observed. It became clear that controllers tuned at a single-fixed operating point are not satisfactory if the plant has to operate at different conditions.

The prevailing technology is PID control. The controllers are tuned at *some* operating point and left. The effects of the controllers, which become detuned at other operating conditions, are ignored until or unless the consequences are *severe*. The problem is one of controlling a *nonlinear plant* for which sophisticated techniques could be attempted against considerable resistance in the prevailing conditions of practice. The prospects of the impact of such techniques on the prevailing practice in MSF plant control are bleak, and the likelihood of such efforts remaining as an academic exercise is high at present.

The approach taken in this study is to accept the prevailing framework of PID control and to enhance it by feasible methods. The objective is to tune the controllers automatically at any point in the operating region. The approach is based on simulation-based optimal design.

At certain points in the operating region, the optimal PID controller parameters have been determined by employing a specially developed simulation facility. For the 155 dimensional model encountered here, methods of controller design based on finite dimensional approaches are neither easy nor practically sensible because of the irreducibility of the model to a tractable size. The facility developed here uses the plant model in nonparametric form such as a step response, which is more easily obtainable model in practice by simple experiment than the parametric counterparts, which call for sophisticated methods of identification. This seems to be the practical reason for the ubiquity of PID controllers and lack of finite dimensional approaches to control design in the process industry.

With the operating region sampled at 12 points (12 cases), and the optimal controller parameters obtained at each of these, the possibility for a simple adaptive control strategy was created. Adaptive control by parameter scheduling has been proposed in two ways using the information at these 12 points. In one approach, an adaptive parameter scheduling law has been suggested in terms of a mapping method. In the other, an ANN has been trained with the information from these 12 points to model a nonlinear relationship rather than a simple linear one between the set of optimal controller parameters and the set of operating conditions (recycle flow, TBT). The results by both methods are comparable. It should be remembered, however, that the ANN is capable of modeling complex surfaces, not mere planes.

The book also discussed certain issues related to practical plant measurement and control. An integrated framework for a practical implementation of measurement, data reconciliation, modeling, and control has been proposed. The work has thus attempted to suggest enhancements in the existing practices by feasible and affordable extensions in the control of large MSF plants. Looking back, it has been a considerable effort with a satisfactory improvement. This modest contribution obviously has to be followed by much further work as indicated here.

9.2 Suggested Directions for Further Work

There are several aspects of MSF plants that need further investigation. The foremost among them is the plant model itself, which requires a systematic refinement. Comprehensive investigations on fouling, brine orifice models, and venting phenomena have to be incorporated in the model. This task is by no means simple. It calls for a full-scale investigation of these phenomena.

The operating conditions are affected by the fouling degree of evaporator and the estimation of the fouling factors is necessary at predetermined time intervals, in order to adapt the mathematical model to the actual conditions at any time, since fouling develops gradually in time and makes the model correspondingly drift in its character. The process of modeling becomes complicated by the so-called ball cleaning method, which is employed to combat fouling. Likewise, the state of modeling of the interstage orifices is far from satisfactory. This calls for detailed investigations.

Online parameter estimation may be an alternative to estimate the model parameters using plant measurements. The use of such methods in nonlinear models calls for special techniques. Any result that is based on modeling relies heavily on the quality of the model itself. Control strategies depend on how the models are characterized. For instance, if estimation aims at both the model and the associated uncertainty, the stage is set for the possibility of robust control.

An overall steady-state optimization of the process can be done, using this model to determine desirable operating conditions. A choice is possible among different objective functions, either technical or economical. At the same time, process constraints as well as the inequality constraints on certain operating parameters must be satisfied. The model equations, constraints, and objective function as well are highly nonlinear. Therefore, proper choice of optimization technique is crucial.

In conclusion, it should be recognized that efforts of single individuals would hardly suffice to tackle the large and complex modeling problem with several issues of concern. Coherent and well-coordinated teamwork is the only approach to the solution of the several problems in MSF desalination process control. The nature of the problem is interdisciplinary. On the surface, the process may seem to be simple—merely as one of heating, evaporation, and condensation, but it is energy intensive and thus expensive at the scales of production required to supply water in place of natural resources of freshwater. Therefore, MSF desalination processes, which support life in many parts of the world, deserve much research attention and it is hoped that they will receive their due in the future.

The final goal of any studies such as those presented here is to make the system optimally operational in the sense of efficient utilization of precious natural resources—such as energy. With the MSF process being energy-intensive, alternative energy sources must replace the present heat coming from the LP steam of power plants, which presently run on fossil fuel. Large amounts of potable water can be obtained by using thermal desalination in cogeneration plants in which the LP steam is used as the energy source to heat seawater. The availability of freshwater is thus linked to the availability of energy. The scarcity of water is acute in the gulf region and as long as the fossil energy sources exist, the present method of seawater desalination is viable but not sustainable in the long run. For the desalination process to be sustainable, the energy must be from a renewable source.

Fortunately, the Arab region is endowed with abundant solar energy coming from above free of charge as is the case with some other parts of the world. This energy must be put to use in desalinating seawater in this region. There are thoughts around the world on the possibility of solar energy plants in the solar belt of the world, which covers much of the Arab land. It is hoped that these plants may be connected in a grid, and excess energy be supplied to the rest of the world. As far as the link between solar energy potential in the Arab world and desalination is concerned, the following estimates are noteworthy (Al Gobaisi, 2010):

- Total solar thermal energy falling on the Arab land is 28.623 million TWh/year or 17,889 billion barrels of oil equivalent (BOE) annually.
- This *annual recurring (renewable)* amount is about 27.515.2 times the *total* existing Arab oil reserves (650 billion barrels), which are *nonrenewable.*
- If we roughly assume that the water deficit in the region by 2050 is likely to be equivalent to 150 km^3 and solar-powered desalination will be used to produce this quantity, then the energy needed will be about = 150×10^9 m^3 $\times$ 5 kWh/m^3 = 750 TWh/year.
- The total electric energy demand for the Arab region is expected to be about 2830 TWh/year in 2050, assuming a reasonable growth of about 3.5% annually.
- The total energy (2830 + 750) (TWh/year) for meeting power and water demand by 2050 will require an area of about 10,156 km^2, which is 0.0725% of the Arab region surface area.
- One square kilometer of the solar collector area can also produce desalinated water of about 193,150 m^3 daily or 70.5 million m^3 annually, assuming an average electric energy consumption for desalinated water to be about 5 kWh/m^3.
- The electrical energy required to produce a desalinated water equivalent to the total water flowing in all the rivers (Nile = 84 km^3/year, Euphrates = 30 km^3/year, and Tigris = 21.2 km^3/year) = 135×10^9 m^3/year $\times$ 5 kWh/m^3 = 675 TWh/year.
- The total solar collector surface area required to generate this amount of energy is about 1915 km^2.

Thus, a solar collector of 1915 km^2 can produce *three more* rivers (Nile, Euphrates, and Tigris) of desalinated water!

Solar energy plants are impressively on the rise in recent years all over the world. The United Arab Emirates has three to mention. Two Al Shams plants and the Masdar City plant have attained global recognition. The latter is a model green city, which is entirely powered by solar energy.

Sustainability of energy and water systems rests on the use of energy and water sources with minimal ecological footprint.

First, solar energy plants must also be considered for land use, which affects the environment. Likewise, desalination plants that derive their input water primarily from water bodies such as the sea must minimize their impact on the marine life in the source, which is due to effluent discharges that cause both chemical and thermal pollution.

In future decades, the increasing potable water scarcity in the world will give rise to more and more desalination plants, and a majority of these will be MSF units on a large scale. Although in terms of physics, the process is similar to what takes place in the natural hydrological cycle that is powered by the sun and simple to understand, the engineering of MSF plants needs to be supported by detailed studies, efficient design, and management to ensure sustainability. Let us hope that desalination in general and MSF processes in particular will take center stage in academic and research institutions around the world toward providing sustainable solutions to both the energy and water problems of global society. Let us also hope that desalination plants will in the future be powered by solar energy or operate in cogeneration with solar power plants. Of course, in such a case, solar energy plants have to be near the sea.

Reference

Al Gobaisi, D.M.K.F., B. Makkawi, and A.M El-Nashar (2010), Renewable energy versus nuclear energy, in *International Conference on Renewable and Alternative Sources of Energy*, Beirut, Lebanon, November 25–26.

10

Programs

10.1 SPEEDUP Package

SPEEDUP is an equation-based flowsheet simulation package designed to do the following:

- Solve steady-state process simulation and design problems.
- Optimize steady-state solutions using an objective function and constraints supplied by the user.
- Simulate dynamic processes where the variables change with time. The model will contain differential equations, and operational variables may be defined as functions of time.
- Model the dynamics of control systems for new designs or for tuning existing control loops.
- Fit model parameters to the experimental data.
- Interface with the operating processes for online optimization.

SPEEDUP will also accomplish the following:

- Display and print results in graphical or tabular form.
- Operate from a library of steady-state and dynamic unit operation models.
- Provide a modeling capability, which allows users to add models for new unit operations by entering the model equations. The user does not need to know the order in which the equations should be solved and no knowledge of computer programming is required.
- Allow existing programs for unit operations or physical property and thermodynamic calculations to be interfaced.

The MSF process model equations consist of mass and energy balance equations for all sections in the MSF plant together with the associated

correlations for heat transfer and physical properties. Each process stream
has four attributes defining the stream, namely, flowrate, temperature, concentration, and enthalpy.

10.1.1 Problem Description

The problem to be solved is described in the SPEEDUP input language,
which consists of a number of input sections, each of which describes a specific aspect of the problem. Any given problem description is unlikely to use
all types of sections. The sections used will depend upon the nature of the
problem. These are classified as follows.
  Sections required for all simulations include the following:

  UNIT: Specify a model, such as FLASH or PUMP, for each unit in the
  flowsheet
  FLOWSHEET: Indicate connections between the units as well as any
  information flows such as controller action
  DECLARE: Give initial values and allowable ranges for types of variables and define the stream structures
  OPERATION: Enter the operational data and specifications

Optional sections include the following:

  OPTIONS: Change the default values for calculation options such as
  tolerances and print levels
  TITLE: Enter a title for the simulation to appear on the output
  GLOBAL: Impose general constraints (specifications) on the process

For optimization:

  GLOBAL: Define the objective function and the constraints

For dynamic simulation:

  CONDITIONS: Print information and warnings or terminate the calculation when certain conditions are met

When defining user-written models:

  MODEL: Write the model description
  PROCEDURE: Give any FORTRAN subroutines, which are to be interfaced to SPEEDUP

10.1.2 Database

Before a problem can be run, the problem description must be stored on the SPEEDUP database. More than one problem description may be held on the database at any time, and problems remain there until they are specifically deleted or the database file is erased.

A problem description may be created, using an editor, in a data file outside of SPEEDUP, and then stored into the SPEEDUP database. Alternatively, it may be created section by section within the SPEEDUP Executive.

10.1.3 Executive

When SPEEDUP is invoked it will be taken into the Executive. In this mode, SPEEDUP accepts interactive commands. Whenever any action is completed, it will be returned to the Executive. Being in the Executive will be indicated by the following prompt:

```
Enter command >
```

Within the Executive, commands can be issued to create or edit a problem description, run problems, invoke diagnostic facilities, and display results.

10.1.4 Translator

Translator analyzes the problem description section by section and stores the information in a coded form in the database.

It then creates a FORTRAN program that will be run during the numerical solution phase to solve the problem described in the input file.

10.1.5 Structure of a Model

A SPEEDUP model consists of up to seven subsections:

- Model name
- Help
- Set
- Type
- Stream
- Equation
- Procedure

In this section, the major submodels representing the process are outlined in the following:

Model	Maximum Flow (t/h)
BL_VALVE	5800
COND_VALVE	205
CUL_VALVE	3700
DL_VALVE	1500
DSUP_VALVE	
MKUP_VALVE	7300
RE1_VALVE	15,950
RE2_VALVE	5800
REJ_VALVE	9000
ST_VALVE	20

10.1.5.1 Flash Model

This represents the dynamics of the flash process in flash chamber, cooling tubes and distillate trays and vapor space combined. It accounts for the gas generated (CO_2) and interstage vapor flow and computes individual temperature losses instead of a fixed total loss. A typical flash unit is schematically shown in Figure 10.1 and consists of four portions, namely, a flash chamber, distillate tray, vapor space, and the tube bundle containing

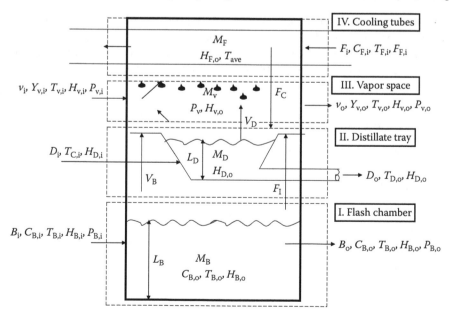

FIGURE 10.1
Flash stage.

cooling brine. The attributes of each stream entering and leaving the flash unit are its flowrate (B, D, V and F, respectively), temperature (T), concentration (C or Y), pressure (P), and specific enthalpy (H). Each portion has a mass holdup denoted by M. Different versions of the flash chamber schematic are given to provide as detailed information as possible including the manner in which the variables are called in the program for the benefit of the reader.

In order to obtain a computationally manageable model of the flash, it is assumed that the flashing brine, vapor, and distillate are well mixed, and the two remain at equilibrium conditions. Hence, the model equations for mass, component, and enthalpy balances for the loops I, II, and III, shown in Figure 10.1, describe the flashing and condensation phenomena occurring in the flash unit. Figures 10.2 through 10.4 show the first stage, a general intermediate stage, and the last stage of the cascade of flash chambers, respectively, indicating the names of the variables that are used in the program.

Loop I:

$$\frac{d(MB)}{dt} = B_i - B_o - VB \tag{10.1}$$

$$\frac{d(MB \cdot CB_o)}{dt} = B_i \cdot CB_i - B_o \cdot CB_o \tag{10.2}$$

$$\frac{d(MB \cdot HB_o)}{dt} = B_i \cdot HB_i - B_o \cdot HB_o - VB \cdot HVB \tag{10.3}$$

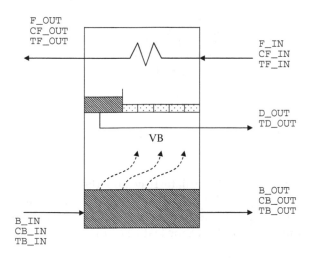

FIGURE 10.2
First flash chamber.

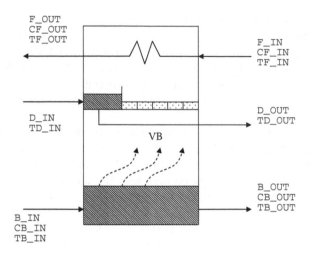

FIGURE 10.3
A general stage in an MSF plant.

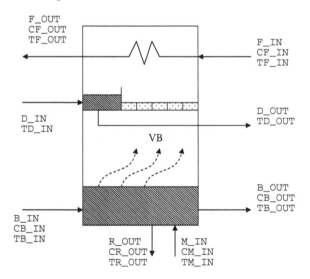

FIGURE 10.4
The last stage in an MSF plant.

Loop II:

This loop consists of the "distillate tray and the vapor space combined." The total mass holdup is $M = MV + MD$. A part of the vapor holdup MV is occupied by MI, being the holdup of inert gases either generated in the flash chamber or entering with the vapor leak. The model equations are

$$\frac{dM}{dt} = D_i + VB + V_i + FI - D_o - V_o \qquad (10.4)$$

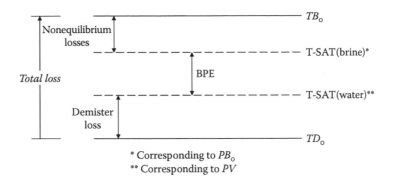

FIGURE 10.5
Temperature losses in the flash.

$$\frac{d(MI)}{dt} = V_i \cdot YI_i + FI - V_o \cdot YI_o \tag{10.5}$$

$$\frac{d(MV \cdot HVD + MD \cdot HD)}{dt} = D_i \cdot HD_i + VB \cdot HVB + V_i \cdot HV_i$$
$$- D_o \cdot HD_o - V_o \cdot HVD + FI \cdot HVB - Q \tag{10.6}$$

The temperature losses that occur between the flash chamber and the vapor space are indicated in Figure 10.5.

Loop III:

The dynamic change in the coolant holdup MW inside the tubes is neglected. The model equations are

$$F_i \, CF_i = F_o \, CF_o \tag{10.7}$$

$$MW \cdot \frac{d(HF_o)}{dt} = F_i(HF_i - HF_o) + Q \tag{10.8}$$

$$Q = U \cdot AHX(TD_o - T_{ave}) \tag{10.9}$$

$$T_{ave} = \frac{(TF_i + TF_o)}{2} \tag{10.10}$$

10.1.5.2 Orifice Model

Interstage flow of the flashing brine occurs through an orifice, which is schematically shown in Figure 10.6. The flow model is developed starting from

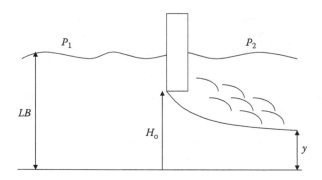

FIGURE 10.6
Interstage transfer through the orifice.

the following equation of energy balance consisting of pressure, potential, and kinetic energy terms:

$$\frac{P_1 + \rho g LB}{\rho} + \frac{v_1^2}{2} = \frac{P_2 + \rho g y}{\rho} + \frac{v_2^2}{2} \tag{10.11}$$

The continuity equation is as follows:

$$v_1 \cdot LB \cdot WB = v_2 \cdot CH \cdot CC \cdot WO \cdot HO \tag{10.12}$$

Substituting Equation 10.12 into 10.11,

$$v_1 = \sqrt{\frac{2\Delta P}{\rho \left(\dfrac{LB^2 \cdot WB^2}{CH^2 \cdot CC^2 \cdot WO^2 \cdot HO^2} - 1 \right)}} \tag{10.13}$$

where

$$\Delta P = P_1 - P_2 + \rho g \left(LB - CC \cdot HO \right) \tag{10.14}$$

Now, the flowrate through the orifice is written as

$$B_o = \rho \cdot LB \cdot WB \cdot v_1 \tag{10.15}$$

Substituting for v_1 from Equation 10.13,

$$B_o = \frac{\left(LB \cdot WB \cdot \sqrt{2\rho\Delta P} \right)}{\sqrt{\left(\dfrac{LB^2 \cdot WB^2}{CH^2 \cdot CC^2 \cdot WO^2 \cdot HO^2} \right) - 1}} \tag{10.16}$$

Multiplying the numerator and denominator of the left-hand side of Equation 10.16

by

$$B_o = WO \cdot HO \cdot CH \sqrt{2\rho \Delta P} \tag{10.17}$$

where

$$CD = \frac{CH \cdot CC}{\sqrt{1 - \left(\dfrac{CH^2 \cdot CC^2 \cdot WO^2 \cdot HO^2}{LB^2 \cdot WB^2} \right)}} \tag{10.18}$$

10.1.5.3 Brine Heater Model

The model equations consist of mass and energy balances, the latter includes a dynamic change in temperature of brine inside the heater transfer tubes (Figure 10.7).

Shell side:

$$\frac{d(MS + MC)}{dt} = FS - FC \tag{10.19}$$

$$\frac{d(MS \cdot HS + MC \cdot HC + MHX \cdot CP_{steel} \cdot TS)}{dt} = FS \cdot HS_i - FC \cdot HC_o - Q \tag{10.20}$$

MC is calculated as a product of a condensate sump cross-sectional area, condensate level, and its density.

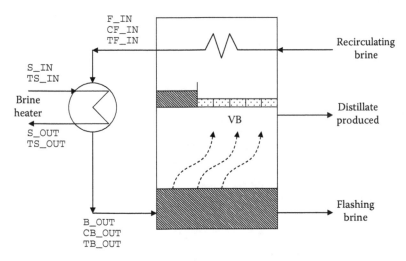

FIGURE 10.7
Brine heater section.

For better representation, the tube side is divided into 10 segments, each segment is described by a lumped model:

$$Q = \sum_{i=1}^{10} Q_i \tag{10.21}$$

where heat transfer per segment

$$Q_i = \frac{U \cdot AHX}{10} \cdot \left(TS - T_{\text{ave},i}\right) \quad i = 1, 2, \ldots, 10 \tag{10.22}$$

$$\left(\frac{MW}{10}\right) \cdot \frac{\mathrm{d}HF_{i+1}}{\mathrm{d}t} = F_i\left(HF_i - HF_{i+1}\right) + Q_i \quad i = 1, 2, \ldots, 10 \tag{10.23}$$

10.1.5.4 Seawater Flow Scheme

The seawater flow schemes in the summer and winter seasons are shown in Figures 10.8 through 10.13, respectively. The same is included in the MSF model by defining four SPLIT models and one MIX model (as indicated in Figures 10.8 and 10.5).

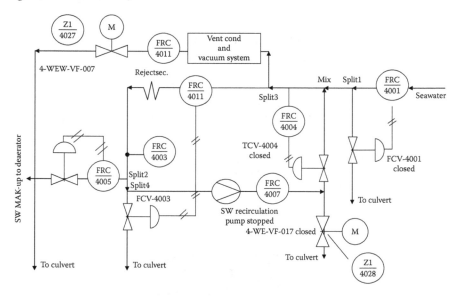

FIGURE 10.8
Seawater flow scheme in the summer season.

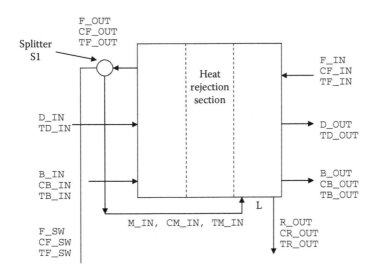

FIGURE 10.9
Splitting point in an MSF plant (summer operation).

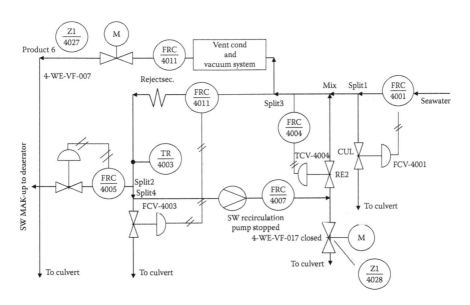

FIGURE 10.10
Seawater flow scheme in the winter season.

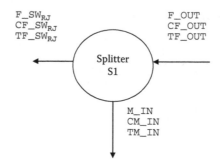

FIGURE 10.11
Splitter.

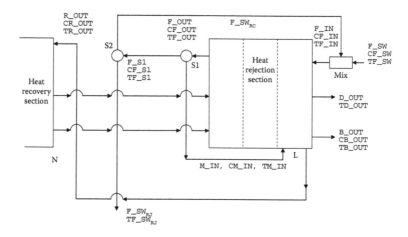

FIGURE 10.12
Splitting and mixing points in an MSF plant (winter operation).

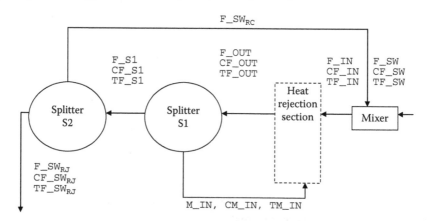

FIGURE 10.13
Splitters and mixer in an MSF plant (winter operation).

10.1.5.5 Dosing Pumps

The following information is provided to introduce the various doses of chemicals required.

Antiscale: (Belgard)

TBT (°C)	PPM (Based on Makeup Flow)
90	1.5–2.0
95	2.0–2.5
100	2.5–3.0
106	3.0–3.5

Hence, the dosing range is set to 0–0.5 L/min.

Lime/Caustic Soda
30 ppm is needed based on the distillate flowrate. Hence, the dosing range is set to 0–1 L/min.

Sodium Sulfite
The dosing range is set to 0–1 L/min.

Nomenclature

AHX	Brine heater, heat transfer area
B	Flashing brine flowrate
CB	Salt concentration in flashing brine
CC	Contract coefficient of orifice
CD	Discharge coefficient of orifice
CF	Salt concentration in cooling brine
CH	Contract coefficient of orifice (horizontal direction)
CP	Specific heat
D	Distillate flowrate
F	Cooling brine flowrate
FI	Inerts flowrate
FS	Steam flowrate
g	Gravitational constant
HB	Specific enthalpy of flashing brine
HC	Specific enthalpy of steam condensate
HD	Specific enthalpy of distillate
HF	Specific enthalpy of cooling brine
HO	Orifice height
HS	Specific enthalpy of steam
HV	Specific enthalpy of vapor leak
HVB	Specific enthalpy of vapor generated
LB	Brine level
LD	Distillate level

MB	Mass holdup, flash chamber
MC	Mass holdup, steam condensate
MD	Mass holdup, distillate tray
MHX	Mass of steel in brine heater
MI	Mass holdup, inerts
MS	Mass holdup, steam
MV	Mass holdup, vapor
MW	Mass holdup, cooling brine
PB	Pressure, flash chamber
PV	pressure, vapor space
Q	Rate of heat transfer, brine heater
TB	Temperature, flashing brine
TD	Temperature, distillate
TF	Temperature, cooling brine
TS	Temperature, steam
TV	Temperature, vapor
U	Heat transfer coefficient
V	Vapor flowrate
v	Linear velocity
VB	Flowrate of vapor generated
WB	Width of flash chamber
WO	Width of orifice
YI	Mass fraction inert gases
ρ	Density
i	Inlet
o	Outlet

10.1.6 SPEEDUP Simulation Program List

```
OPTIONS
ROUTINES NEWTON, SUPERDAE
EXECUTION
PRINT_LEVEL = 0
TARGET = TERMINAL
HIRESIDUAL = 4
HIVARSTEPS = 5
SS_TOL = 1E-6
FDPERTURB = 1E-6
TIME_STEP = 0.1
INTERVALS = 800 #120
BOUNDCHECK = OFF
****
DECLARE
TYPE
FLOWRATE = 199:0:2727 UNIT = "t/min"
DOSING_FLOW = 0: 0:1000 UNIT = "l/min"
```

```
VAP_FLOW = 1:0.0:27 UNIT = "t/min"
DIST_FLOW = 1:0.27:45 UNIT = "t/min"
TEMPERATURE = 33:1: 260 UNIT = "c"
TEMP_DIFF = 8:0.0: 83 UNIT = "c"
TEMP_LOSS = 0.35: 0.005: 1.0 UNIT = "C"
ENTHALPY = 70:15:400 UNIT = "kcal/kg"
ENTH_FLOW = 2E5:0:5E6 UNIT = "KCAL/MIN"
VAP_ENTH = 555:400:1111 UNIT = "kcal/kg"
CONCENTRATION = 0.05:0.00:0.1 UNIT = "mass frac"
HEAT_CAPACITY = 1: 0.1: 2 UNIT = "kcal/kg-C"
AREA = 45:0.1:1E4 UNIT = "m2"
S_AREA = 3000:92:1E4 UNIT = "m2"
D_AREA = 3:2.0: 6.0 UNIT = "m2"
LENGTH = 0.6:0.0: 25 UNIT = "m"
VOLUME = 28:0.1:1000 UNIT = "m3"
MASS = 28:0.0001:454 UNIT = "t"
KG_MASS = 100:0.001:1E6 UNIT = "kg"
ENERGY = 100:0:1E7 UNIT = "kcal"
OHTC = 24:15:70 UNIT = "kcal/min-m2-C"
DENSITY = 1000:0.01:1620 UNIT = "kg/m3"
DIAMETER = 24: 12:508 UNIT = "mm"
TUBE_ID = 24:1:500 UNIT = "mm"
RATIO = 0.5:0:1
NOTYPE = 52: -1E9:1E9
TUBE_AREA = 3e-4:9e-6:1 UNIT = "m2"
TUBE_LENGTH = 15:6:25 UNIT = "m"
TUBE_VOLUME = 0.005:3E-5:0.03 UNIT = "m3"
LEVEL = 0.6:1E-6:1.5 UNIT = "m"
PRESSURE = 0.5:0.001:2 UNIT = "bar"
FOULING_FACTOR = 0.01:0.0:0.025 UNIT = "hr-m2-c/kcal"
TUBE_MASS = 30e3:10e3:40e3 UNIT = "kg"
POSITIVE 0.5:0:10
DELTA_P = 0.15:1e-5: 10 UNIT = "BAR"
THICKNESS = 1:0:50 UNIT = "mm"
MOLE_FRACTION = 0.1:0:1
MASS_FRACTION = 0.1:0:1
MOL_WEIGHT = 10:1:500       UNIT = "kg/mol"
```

STREAM MAINSTREAM
```
TYPE
FLOWRATE, TEMPERATURE, CONCENTRATION, ENTHALPY
****
```
STREAM BRINE
```
TYPE
FLOWRATE, TEMPERATURE, PRESSURE, CONCENTRATION, ENTHALPY
****
```
STREAM DISTILLATE
```
TYPE
DIST_FLOW, TEMPERATURE, ENTHALPY
****
```

```
STREAM STEAM
TYPE
VAP_FLOW, TEMPERATURE, PRESSURE, VAP_ENTH
****
STREAM CONDENSATE
TYPE
FLOWRATE, TEMPERATURE, PRESSURE, ENTHALPY
****
STREAM VAPOUR
TYPE
VAP_FLOW, TEMPERATURE, PRESSURE, MASS_FRACTION, VAP_ENTH
****
PROCEDURE BPE
INPUT
TEMPERATURE, CONCENTRATION
OUTPUT
TEMP_DIFF
****
PROCEDURE BRH
INPUT
CONCENTRATION, TEMPERATURE
OUTPUT
ENTHALPY
****
PROCEDURE BRHI
INPUT
CONCENTRATION, TEMPERATURE (NSLICES)
OUTPUT
ENTHALPY (NSLICES)
****
PROCEDURE BRINCP
INPUT
CONCENTRATION, TEMPERATURE
OUTPUT
HEAT_CAPACITY
****
PROCEDURE BRINEH
INPUT
CONCENTRATION, TEMPERATURE
OUTPUT
ENTHALPY
****
PROCEDURE BRINRO
INPUT
CONCENTRATION, TEMPERATURE
OUTPUT
DENSITY
****
```

PROCEDURE EXX
INPUT
TEMPERATURE, TEMPERATURE, TEMPERATURE, TUBE_LENGTH,
LEVEL, FLOWRATE
OUTPUT
TEMP_DIFF
****
PROCEDURE HCOEFF
INPUT
TEMPERATURE, TEMPERATURE, CONCENTRATION, TUBE_ID,
TUBE_LENGIH, FLOWRATE, AREA, FOULING_FACTOR
OUTPUT
OHTC
****
PROCEDURE HTCRIG
INPUT
FLOWRATE, TEMPERATURE, CONCENTRATION, FLOWRATE,
TEMPERATURE, FOULING_FACTOR, TUBE_ID, TUBE_LENGTH, THICKNESS,
NOTYPE
OUTPUT
OHTC
****
PROCEDURE ORIFCD
INPUT
LENGTH, LENGTH, LENGTH, LEVEL, PRESSURE, PRESSURE,
DENSITY
OUTPUT
POSITIVE
****
PROCEDURE PVAP
INPUT
TEMPERATURE, CONCENTRATION
OUTPUT
PRESSURE
****
PROCEDURE STEAMH
INPUT
TEMPERATURE
OUTPUT
VAP_ENTH
****
PROCEDURE STEAMP
INPUT
TEMPERATURE
OUTPUT
PRESSURE
****
PROCEDURE STMH
INPUT
TEMPERATURE

```
OUTPUT
VAP_ENTH
****
```

PROCEDURE TEXX
```
INPUT
REAL, TEMPERATURE, TEMPERATURE
OUTPUT
TEMP_DIFF
****
```

PROCEDURE TLOSS
```
INPUT
TEMPERATURE
OUTPUT
TEMP_DIFF
```

PROCEDURE WATERH
```
INPUT
TEMPERATURE
OUTPUT
ENTHALPY
****
```

PROCEDURE WATERO
```
INPUT
TEMPERATURE
OUTPUT
DENSITY
****
```

MODEL BRINE_HEATER
SET
```
STM_Z = 1.12,
R_GAS = 0.08314,
TON_TO_KG = 1000,
C_TO_K = 273.15,
CP_STEEL = 0.12,
COND_TC = 1.0,
NSLICES = 10,
NSLICES1 = 9
TYPE
#Streams
B_IN, B_OUT AS FLOWRATE
CB_IN, CB_OUT AS CONCENTRATION
TB_IN, TB_OUT AS TEMPERATURE
HB_IN, HB_OUT AS ENTHALPY
FS, FC AS FLOWRATE
TS_IN, TC_OUT AS TEMPERATURE
PS_IN, PC_OUT AS PRESSURE
HS_IN AS VAP_ENTH
HC_OUT AS ENTHALPY
#Connections
TB_sig AS TEMPERATURE #of brine leaving exchanger
LC AS LEVEL #of condensate
```

```
# Set Variables
*V AS VOLUME #of shell
*RHO_C AS DENSITY #of condensate
*AHX AS AREA #for heat exchange
*MB AS KG_MASS #of brine in tubes, kg
*ASUMP AS AREA#cross section, of condensate sump
*SIG_DEL AS NOTYPE #delay on temperature signal
*TUBE_LENGTH AS TUBE_LENGTH
*TUBE_ID AS TUBE_ID
*FOULING AS FOULING_FACTOR
*MHX AS KG_MASS #of metal in heat exchanger
*ZERO_CONC AS CONCENTRATION #*MUST* be zero in OPERAT
*N_TUBE AS NOTYPE
*TUBE_THICKNESS AS THICKNESS
#Internal Variables
M AS KG_MASS #of steam+condensate
E AS ENERGY#total, of system
Q AS ENTH_FLOW #exchanger duty
QI AS ARRAY (NSLICES) OF ENTH_FLOW#duty, for each slice
T_AVI AS ARRAY (NSLICES) OF TEMPERATURE #average, of brine
TBI AS ARRAY(NSLICES)OF TEMPERATURE#outlet of each slice
HBI AS ARRAY(NSLICES)OF ENTHALPY #outlet, of each slice
MS AS KG_MASS #of steam
MC AS KG_MASS #of condensate
HS AS VAP_ENTH #of condensing steam
HC AS ENTHALPY #of condensate
RHO_AS AS DENSITY #of condensing steam
TS AS TEMPERATURE #of condensing steam
PS AS PRESSURE #inside shell
U AS OHTC #overall htc for exchanger
U_hr AS NOTYPE #ohtc per hour
T_AV AS TEMPERATURE
FCOND AS FLOWRATE

STREAM
INPUT 1 B_IN, TB_IN, CB_IN, HB_IN
OUTPUT 1 B_OUT, TB_OUT, CB_OUT, HB_OUT
INPUT 2 FS, TS_IN, PS_IN, HS_IN
OUTPUT 2 FC, TC_OUT, PC_OUT, HC_OUT

CONNECTION 1 TB_sig
CONNECTION 2 LC

EQUATION
#
#Shell Side
#

#Mass Balance
$M = TON_TO_KG*(FS-FC);
```

```
#Energy Balance
$E = TON_TO_KG*(FS*HS_IN-FC*HC_OUT)-Q;

#Mass Constraint
M = MS+MC;

#Energy Constraint
E = MS*HS+MC*HC+MHX*CP_STEEL*TS;

#Volume constraint

V = MS/RHO_S+MC/RHO_C;

#Condensate level

MC = ASUMP*LC*RHO_C;

#Heat Duty
Q = SIGMA (QI);
#Tube Side

#Average Temperature
T_AV (1) = 0.5*(TB_IN+TBI(1));
T_AVI(2:NSLICES) = 0.5*(TBI(1:NSLICES1)+TBI(2 :NSLICES));

#Heat flow per slice
QI (1 :NSLICES) = U*AHX/NSLICES*(TS-T_AVI)(1 :NSLICES));

#Brine Energy Balance

MB/NSLICES*$HBI (1) = TON_TO_KG*B_IN* (HB_IN-HBI(1))+QI(1);
MB/NSLICES*$HBI(2 :NSLICES) = TON_TO_KG*B_IN*
(HBI(1:NSLICES1)-HBI(2:NSLICES))+QI(2:NSLICES);

#Flow constraint
B_OUT = B_IN;

#Concentration constraint
CB_OUT = CB_IN;

#General Equalities

HB_OUT = HBI(NSLICES);
TC_OUT = TS;
PS = PC_OUT = PS_IN;
TB_OUT = TBI(NSLICES);
TB_sig = DELAY TB_OUT BY SIG_DEL;
HC_OUT = HC;
T_AV = SIGMA (T_AVI)/NSLICES;
U_hr = U*60;

#Thermophysical Properties

#Steam density
RHO_S = (18.0*PS)/(STM_Z*R_GAS*(TS+C_TO_K));
```

```
#Mean condensate flow (steam flow with a lag)
COND_TC*$FCOND+FCOND = FS;


PROCEDURE

#Saturation pressure
(PS) PVAP (TS, ZERO_CONC)

#Steam enthalpy
(HS) STMH (TS)

#Condensate enthalpy
(HC) WATERH (TS)

#Brine enthalpy
(HBI) BRHI (CB_OUT, TBI)

#Overall Heat Transfer Coefficient
(U) HTCRIG(B_IN, T_AV, CB_IN, FCOND, TS, FOULING,
TUBE_ID, TUBE_LENGTH, TUBE_THICKNESS, N_TUBE)
```

#Steam Desuperheater

```
MODEL DESUP
TYPE
#Inputs
S_INA S FLOWRATE
TS_IN AS TEMPERATURE
PS_IN AS PRESSURE
HS_IN AS VAP_ENTH
C_TN AS FLOWRATE
TC_IN AS TEMPERATURE
PC_IN AS PRESSURE
HC_IN AS ENTHALPY
#Output
S_OUT AS FLOWRATE
TS_OUT AS TEMPERATURE
PS_OUT AS PRESSURE
HS_OUT AS VAP_ENTH
#Connection
T_SIG AS TEMPERATURE

#Internal variables
HS AS VAP_ENTH
HC AS ENTHALPY
Vf AS NOTYPE
T_SAT AS TEMPERATURE
T_SUP AS TEMPERATURE
*C_CZERO AS CONCENTRATION #must be set to zero!
```

```
STREAM
INPUT 1 S_IN, TS_IN, PS_IN, HS_IN
INPUT 2 C_IN, TC_IN, PC_IN, HC_IN
OUTPUT 1 S_OUT, TS_OUT, PS_OUT, HS_OUT
CONNECTION 1 T_SIG
EQUATION
#Mass balance
S_OUT = S_IN+C_IN
#Enthalpy balance
HS_OUT*S_OUT = HS_IN*S_IN+HC_IN*C_IN;
#Pressure equality
PS_OUT = PS_IN = PC_IN;
#Vapour fraction at outlet
HS_OUT = vf*HS+(1-vf)*HC;
#Outlet temperature
IF vf > 1 THEN
TS_OUT = T_SUP
ELSE
TS_OUT = T_SAT
ENDIF;
#Signal temperature
T_SIG = TS_OUT;
PROCEDURE
#Steam enthalpy
(HS) STMH (T_SAT)
#Condensate Enthalpy
(HC) WATERH (T_SAT)
#Saturation Temperature
(PS_OUT) PVAP (T_SAT, C_ZERO)
#Superheated Enthalpy
(HS_OUT) STMH (T_SUP)

FLASH_FIRST
SET
G = 9.81456,
TON_TO_KG = 1000,
MW_H2O = 18.0,
C_TO_K = 273.15,
R_GAS = 0.08314,
CRIT = 0.5,
STAGE_NO #you must set this in UNIT for each stage

TYPE
F_IN, F_OUT, B_IN, B_OUT AS FLOWRATE
D_OUTAS DIST_FLOW
TF_IN, TF_OUT, TD_OUT, TB_IN, TB_OUT AS TEMPERATURE
CF_IN, CF_OUT, CB_IN, CB_OUT AS CONCENTRATION
HF_IN, HF_OUT, HB_IN, HB_OUT AS ENTHALPY
HD_OUT AS ENTHALPY
PB_OUT AS PRESSURE
```

```
V_OUT AS VAP_FLOW
TV_OUT AS TEMPERATURE
PV_OUT AS PRESSURE
YI_OUT AS MASS FRACTION
HV_OUT AS VAP_ENTH

#Set variables
*AREA_B AS AREA #x-section, of brine chamber
*AREA_HX AS AREA #surface, of tubes
*AREA_D AS AREA #x-section, of dist tray
*N_TUBE AS NOTYPE
*TUBE_AREA AS TUBE_AREA #x-section, per tube
*TUBE_LENGTH AS TUBE_LENGTH
*DIST_COF AS NOTYPE
*TUBE_ID AS TUBE_ID
*FOULING AS FOULING_FACTOR
*HEIGHT AS LENGTH
*C_ZERO AS CONCENTRATION #must be zero!
*K_DEM AS NOTYPE
*TUBE_THICKNESS AS THICKNESS #mm
*INERTS_RATIO AS NOTYPE
*K_ORIF AS NOTYPE
*MW_INERTS AS MOL_WEIGHT
#Internal variables

V                        AS VOLUME #of flash chamber
PV, PV_SAT               AS PRESSURE
FLOW                     AS FLOWRATE
DISTILLATE               AS DIST_FLOW
HD                       AS ENTHALPY
HVB, HVD, HVBS, HVV      AS VAP_ENTH
HB                       AS ENTHALPY
HF                       AS ENTHALPY.
MB, MD, MV, MI           AS KG_MASS
MW                       AS TUBE_MASS
CB                       AS CONCENTRATION
RHO_B, RHO_F, RHO_D, RHO_V AS DENSITY
LEVEL_D, LEVEL           AS LEVEL
T_STA, TV_SAT            AS TEMPERATURE
T_AVE                    AS TEMPERATURE
VB                       AS VAP_FLOW
UO                       AS OHTC
UO_HR                    AS NOTYPE
VOL_TUBE                 AS TUBE_VOLUME
Q                        AS ENTH_FLOW
TF                       AS TEMPERATURE
T_EXX                    AS TEMP_DIFF
WB                       AS TUBE_LENGTH
DP_DEM                   AS DELTA_P
FC                       AS FLOWRATE
```

```
FI_IN                           AS VAP_FLOW
YI_MOLE                         AS MOLE_FRACTION
P_MIN, P_ORIF                   AS PRESSURE
DP_ORIF                         AS DELTA_P
PB_SAT, PD_SAT                  AS PRESSURE
VD                              AS VAP_FLOW

STREAM
INPUT          1     F_IN, TF_IN, CF_IN, HF_IN
OUTPUT         1     F_OUT, TF_OUT, CF_OUT, HF_OUT
INPUT          2     B_IN, TB_IN, CB_IN, HB_IN
OUTPUT         2     D_OUT, TD_OUT, HD_OUT
OUTPUT         3     B_OUT, TB_OUT, PB_OUT, CB_OUT, HB_OUT
OUTPUT         4     V_OUT, TV_OUT, PV_OUT, YI_OUT, HV_OUT

CONNECTION     1     LEVEL
CONNECTION     2     DISTILLATE
CONNECTION     3     FLOW
CONNECTION     4     LEVEL_D
EQUATION

#Brine balances

#Mass balance flash chamber
$MB = TON_TO_KG*(B_IN-B_OUT -VB);

#Concentration Balance Flash chamber
$MB*CB+$CB*MB = TON_TO_KG *' (B_IN*CB_IN-B_OUT*CB);

#Ethalpy balance of flash chamber
$MB*HB+$HB*MB = TON_TO_KG*(B_IN*HB_IN-VB*HVB-B_OUT*HB);

#Volume/level relationship
MB = RHO_B*AREA_B*LEVEL;

#Assume perfect mixing
HB = HB_OUT;
CB = CB_OUT;
#

#HVB = HVBS+0.48* (TB_OUT-T_SAT);
#
#Non-equilibrium allowance
TB_OUT = T_SAT+T_EXX;

#Partial pressure of inerts
PB_SAT = PB_OUT*(1-YI_MOLE);

#Cooling Tubes

#Mass Balance Tube Side
F_IN = F_OUT = FLOW;
```

```
#Concentration Balance Tube Side
CF_IN = CF_OUT;

#Enthalpy Balance Tubeside
MW*$HF = TON_TO_KG*F_IN*(HF_IN-HF)+Q;

#Heat Transfer
Q = UO*AREA_HX*(TD_OUT-T_AVE);
T_AVE = (TF_IN+TF_OUT)/2;

#Assume Perfect Mixing
HF = HF_OUT;

#Tube Holdup
MW = RHO_F*VOL_TUBE*N_TUBE;
VOL _TUBE - = TUBE_AREA*TUBE_LENGTH;

#Assume Mixing On Tube Side
TF = TF_OUT

#Width of Brine Chamber Is Same As Tube Length
WB = TUBE_LENGTH;

#OHTC in Hours
UO_HR = UO*60;
#Vapour Space

#Rate of inerts Generation
FI_IN = INERTS_RATIO*B_IN;

#Inerts Balance
$MI = TON_TO_KG*(FI_IN-V_OPT*YI_OUT);

#Mass Balance
$MV = TON_TO_KG*(VB+VD-V_ OUT+FI_IN-FC);

#Enthalpy Balance
(VB+FI_IN) *HVB+VD*HVD = V_OUT*HVV+FC*HVV;

#Volume Constraint
V = MV/RHO_V+MD/RHO_D+MB/RHO_B+VOL_TUBE*N_TUBE;

#Mass and mole fraction of inerts
YI_OUT = MI/MV;
YI_MOLE = (MI/MW_INERTS)/((MV-MI)/MW_H2O+MI/MW_INERTS);

#Flow of vapour Extracted
P_MIN = CRIT*PV;
if PV_OUT < P_MIN then
 P_ORIF = P_MIN
```

```
else
 P_ORIF = PV_OUT
endif;
DP_ORIF = PV-P_ORIF;
V_OUT = K_ORIF*SQRT (RHO_V*DP_ORIF);

#Vapour Streams
HVV = HVD+0.48*(TV_OUT-TD_OUT);

HV_OUT = HVV;
PV_SAT = PV*(1-YI_MOLE);

#Demister Losses
DP_DEM = K_DEM/RHO_V*VB*VB;

#Pressure Equality
PV = PB_OUT-DP_DEM;

#Volume of Brine Chamber Is Area*Height
V = AREA_B*HEIGHT;

#Condensation rate (approximately-this must not be negative!)
FC = Q/(HV_OUT-HD)/TON_TO_KG;

#Distillate Balances
#Mass balance
$MD = TON_TO_KG* (FC-D_OUT-VD);
#Enthalpy Balance
FC*HD = D_OUT*HD_OUT+VD*HVD;

#Volume/level relationship
MD = RHO_D*AREA_D*LEVEL_D;

#Assume perfect mixing
HD = HD_OUT;

#Pressure drop equation for distillate flow leaving
D_OUT = DIST_COF*LEVEL_D;

#External connection
D_OUT = DISTILLATE;

#Partial pressure of inerts
PD_SAT = PV*(1-YI_MOLE);

#Property procedures

#Vapour density (neglect inerts)
RHO_V = (MW_H2O*PV)/(R_GAS*(TV_OUT+C_TO_K));
```

```
PROCEDURE
(T_EXX) TEXX (stage_no, TB_IN, TB_OUT)
(PB_SAT) PVAP (T_SAT; CB_OUT)
(PD_SAT) PVAP (TD_OUT, C_ZERO)
(HB) BRH (CB_OUT, TB_OUT)
(HVBS) STMH (T_SAT)
(HVD) STMH (TD_OUT)
(RHO_B) BRINRO (CB_OUT, TB_OUT)
(HF_OUT) BRH (CF_OUT, TF_OUT)
(HD) WATERH (TD_OUT)
(RHO_F) BRINRO (CF_OUT, TF_OUT)
(RHO_D) WATERO (TD_OUT)
(UO) HTCRIG (F_OUT, T_AVE, CF_IN, FC, TD_OUT,FOULING, TUBE_ID,
      TUBE_LENGTH,TUBE_THICKNESS, N_TUBE)
(HVB) VAPH (TB_OUT, PB_OUT)
(HVV) VAPH (TV_OUT, PV)

(PV_SAT) PVAP (TV_SAT, C_ZERO)

MODEL FLASH
SET
G = 9.81456,
TON_TO_KG = 1000;
MW_H2O = 18.0
C_TO_K = 273.15
R_GAS = 0.08314
CRIT = 0.5
STAGE_NO#must be set in UNIT for each stage
TYPE
F_IN, F_OUT, B_IN, B_OUT AS FLOWRATE
D_IN, D_OUT AS DIST_FLOW
TF_IN, TF_OUT, TD_IN, TD_OUT, TB_IN, TB_OUT AS TEMPERATURE
CF_IN, CF_OUT, CB_IN, CB_OUT AS CONCENTRATION
HF_IN, HF_OUT, HB_IN, HB_OUT AS ENTHALPY
HD_IN, HD_OUT AS ENTHALPY
PB_IN, PB_OUT AS PRESSURE
V_IN, V_OUT AS VAP_FLOW
TV_IN, TV_OUT AS TEMPERATURE
PV_IN, PV_OUT AS PRESSURE
YI_IN, YI_OUT AS MASS_FRACTION
HV_IN, HV_OUT AS VAP_ENTH

#Set variables
*AREA_B AS AREA #x-section, of brine chamber
*AREA_HX AS AREA #surface, of tubes
*AREA_D AS AREA #x-sec, of dist tray
*N_TUBE AS NOTYPE
*TUBE_AREA AS TUBE_AREA #x-sec, per tube
*TUBE_LENGTH AS TUBE_LENGTH
```

```
*DIST_COF AS NOTYPE
*TUBE_ID AS TUBE_ID
*FOULING AS FOULING_FACTOR #m2-h-C/kcal
*HEIGHT AS LENGTH
*C_ZERO AS CONCENTRATION #must be zero!
*K_DEM AS NOTYPE
*TUBE_THICKNESS AS THICKNESS #mm
*INERTS_RATIO AS NOTYPE
*K_ORIF AS NOTYPE
*MW_INERTS AS MOL_WEIGHT

#Internal variables
V AS VOLUME #of flash chamber
PV, PV_SAT AS PRESSURE
FLOW AS FLOWRATE
DISTILLATE AS DIST_FLOW
HD AS ENTHALPY
HVB, HVD, HVBS, HVV AS VAP_ENTH
HB AS ENTHALPY
HF AS ENTHALPY
MB, MD, MV, MI AS KG_MASS
MW AS TUBE_MASS
CB AS CONCENTRATION
RHO_B, RHO_F, RHO_D, RHO_V AS DENSITY
LEVEL_D, LEVEL AS LEVEL
T_SAT, TV_SAT AS TEMPERATURE
T_AVE AS TEMPERATURE
VB AS VAP_FLOW
UO AS OHTC #kcal/min-m2-C
UO_HR AS NOTYPE #kcal/hr-m2-C
VOL_TUBE AS TUBE_VOLUME
Q AS NOTYPE
TF AS TEMPERATURE
T_EXX AS TEMP_DIFF
WB AS TUBE_LENGTH
DP_DEM AS DELTA_P
FC AS FLOWRATE
FI_IN AS VAP_FLOW
YI_MOLE AS MOLE_FRACTION
P_MIN, P_ORIF AS PRESSURE
DP_ORIF AS DELTA_P
PB_SAT, PD_SAT AS PRESSURE
VD AS VAP_FLOW

STREAM
INPUT 1 F_IN, TF_IN, CF_IN, HF_IN
OUTPUT 1 F_OUT, TF_OUT, CF_OUT, HF_OUT
INPUT 2 D_IN, TD_IN, HD_IN
OUTPUT 2 D_OUT, TD_OUT, HD_OUT
INPUT 3       B_IN, TB_IN, PB_IN, CB_IN, HB_IN
```

```
OUTPUT 3     B_OUT, TB_OUT, PB_OUT, CB_OUT, HB_OUT
INPUT 4      V_IN, TV_IN, PV_IN, YI_IN, HV_IN
OUTPUT 4 V_OUT, TV_OUT, PV_OUT, YI_OUT, HV OUT

CONNECTION 1 LEVEL
CONNECTION 2 DISTILLATE
CONNECTION 3 FLOW
CONNECTION 4 LEVEL_D

EQUATION

#Brine balance

#Mass balance flash chamber
$MB = TON_TO_KG*(B_IN-B_OUT-VB);

#Concentration Balance Flash chamber
$MB*CB+$CB*MB = TON_TO_KG*(B_IN*CB_IN-B_OUT*CB);

#Enthalpy balance of flash chamber
$MB*HB+$HB*MB = TON_TO_KG*(B_IN*HB_IN-VB*HVB-B_OUT*HB);

#Volume/level relationship
MB = RHO_B*AREA_B*LEVEL;

#Assume perfect mixing
HB = HB_OUT;
CB = CB_OUT;
#
#HVB = HVBS+0.48* (TB_OUT-T_SAT);
#
#Non-equilibrium allowance
TB_OUT = T_SAT+T_EXX;
#Pressure equality
PB_IN = PB_OUT ;

#Partial pressure of inerts
PB_SAT = PB_OUT*(1-YI_MOLE);

#Cooling tubes

#Mass Balance tube side
F_IN = F_OUT = FLOW;

#Concentration Balance tube side
CF_IN = CF_OUT;

#Enthalpy balance tubeside
MW*$HF = TON_TO_KG*F_IN*(HF_IN-HF)+Q;
```

```
#Heat transfer
Q = UO*AREA_HX*(TD_OUT-T_AVE);
T_AVE = (TF_IN+TF_OUT)/2;

#Assume perfect mixing
HF = HF_OUT;

#Tube holdup
MW = RHO_F*VOL_TUBE*N_TUBE;
VOL_TUBE = TUBE_AREA*TUBE_LENGTH;

#Assume mixing on tube side
TF = TF_OUT;

#Width of brine chamber is same as tube length
WB = TUBE_LENGTH;

#OHTC in hours
UO_HR = UO*60;

#Vapour space

#Rate of inerts generation
FI_IN = INERTS_RATIO*B_IN;

#Inerts balance
$MI = TON_TO_KG*(V_IN*YI_IN-V_OUT*YI_OUT+FI_IN);

#Mass Balance
$MV = TON_TO_KG*(VB+VD+V_IN-V_OUT+FI_IN-FC);

#Enthalpy Balance
(VB+FI_IN)*HVB+VD*HVD+V_IN*HV_IN = V_OUT*HVV+FC*HVV;

#Volume constraint
V = MV/RHO_V+MD/RHO_D+MB/RHO_B+VOL_TUBE*N_TUBE;

#Mass and mole fraction of inerts
YI_OUT = MI/MV;
YI_MOLE = (MI/MW_INERTS)/((MV-MI)/MW_H2O+MI/MW_INERTS);

#Flow of vapour extracted
P_MIN = CRIT*PV;
If PV_OUT < P_MIN then
 P_ORIF = P_MIN
else
 P_ORIF = PV_OUT
endif;
DP_ORIF = PV-P_ORIF;
V_OUT = K_ORIF *-SQRT (RHO_V*DP_ORIF);
```

```
#Vapour streams
PV_IN = PV;

HVV = HVD+0.48*(TV_OUT-TD_OUT);

HV_OUT = HVV;

PV_SAT = PV*(1-YI_MOLE);

#Demister losses
DP_DEM = K_DEM/RHO_V*VB*VB;

#Pressure equality
PV = PB_OUT-DP_DEM;

#Volume of brine chamber is area*height
V = AREA_B*HEIGHT;

#Condensation rate {approximately-this must not be negative!}
FC = Q/(HV_OUT-HD)/TON_TO_KG;
```

#Distillate Balances

```
#Mass Balance

$MD = TON_TO_KG*(D_IN-D_OUT-VD+FC);

#Enthalpy balance
D_IN*HD_IN+FC*HD = D_OUT*HD+VD*HVD;

#Volume/level relationship
MD = RHO_D*AREA_D*LEVEL_D;

#Assume perfect mixing
HD = HD_OUT;

#Pressure drop equation for distillate flow leaving
D_OUT = DIST_COF*LEVEL_D;

#External connection
D_OUT = DISTILLATE;

#Partial pressure of inerts
PD_SAT = PV*(1-YI_MOLE);

#Property procedures

#Vapour density (neglect inerts)
RHO_V = (MW_H2O*PV)/(R_GAS*(TV_OUT+C_TO_K));
```

```
PROCEDURE
(T_EXX)TEXX stage_no, TB_IN, TB_OUT)
(PB_SAT) PVAP (T_SAT, CB_OUT)
(PD_SAT) PVAP (TD_OUT, C_ZERO)
(HB) BRH CB_OUT, TB_OUT)
(HVD) STMH (TD_OUT)
(HVBS) STMH (T_SAT)
(RHO_B) BRINRO (CB_OUT, TB_OUT)
(HF_OUT) BRH (CF_OUT, TF_OUT)
(RHO_F) BRINRO (CF_OUT, TF_OUT)
(HD) WATERH (TD_OUT)
(RHO_D) WATERO (TD_OUT)
(UO) HTCRIG (F_OUT, T_AVE, CF_IN, FC, TD_OUT,FOULING, TUBE_ID,
TUBE_LENGTH, TUBE_THICKNESS, N_TUBE)
(HVB) VAPH (TB_OUT, PB_OUT)
(HVV) VAPH (TV_OUT, PV)
(PV_SAT) PVAP (TV_SAT, C_ZERO)


FLASH_LAST
SET
G = 9.81456,
PAS_TO_BAR = 1.0E-5,
TON_TO_KG = 1000,
MW_H2O = 18.0,
C_TO_K = 273.15,
R_GAS = 0.08314,
CR_IT = 0.5,
STAGE_NO #you must set this in UNIT for each stage


TYPE
F_IN, F_OUT, B_IN, B_OUT, M_IN, R_OUT AS FLOWRATE
D_IN, D_OUT AS DIST_FLOW
TF_IN, TF_OUT, TD_IN, TD_OUT, TB_IN, TB_OUT AS TEMPERATURE
TM_IN, TR_OUT AS TEMPERATURE
CF_IN, CF_OUT, CB_IN, CB_OUT AS CONCENTRATION
CM_IN, CR_OUT AS CONCENTRATION
HF_IN, HF_OUT, HB_IN, HB_OUT AS ENTHALPY
HM_IN, HR_OUT AS ENTHALPY
HD_IN, HD_OUT AS ENTHALPY
PB_IN, PB_OUT AS PRESSURE
V_IN, V_OUT AS VAP_FLOW
TV_IN, TV_OUT AS TEMPERATURE
PV_IN, PV_OUT AS PRESSURE
YI_IN, YI_OUT AS MASS_FRACTION
HV_IN, HV_OUT AS VAP_ENTH


#Set variables

*AREA_B AS AREA #x-section, of brine chamber
*AREA_HX AS AREA #surface, of tubes
```

```
*AREA_D AS AREA #x-sec, of dist tray
*N_TUBE AS NOTYPE
*TUBE_AREA AS TUBE_AREA
*TUBE_LENGTH AS TUBE_LENGTH
*TUBE_ID AS TUBE_ID
*FOULING AS FOULING_FACTOR
*HEIGHT AS LENGTH
*C_ZERO AS CONCENTRATION #must be zero!
*K_DEM AS NOTYPE
*TUBE_THICKNESS AS THICKNESS #mm
*INERTS_RATIO AS NOTYPE
*K_ORIF AS NOTYPE
*MW_INERTS AS MOL_WEIGHT

#Internal variables
V AS VOLUME #of flash chamber
PV, PV_SAT AS PRESSURE
FLOW, REC_FLOW AS FLOWRATE
DISTILLATE AS DIST_FLOW
HD AS ENTHALPY
HVB, HVD, HVBS, HVV AS VAP_ENTH
HB AS ENTHALPY
HF AS ENTHALPY
MB, MD, MV, MI AS KG_MASS
MW AS TUBE_MASS
CB AS CONCENTRATION
RHO_B, RHO_F, RHO_D, RHO_V AS DENSITY
LEVEL_D, LEVEL AS LEVEL
T_SAT, TV_SAT AS TEMPERATURE
T_AVE AS TEMPERATURE
VB AS VAP_FLOW
UO AS OHTC
UO_HR AS NOTYPE
VOL_TUBE AS TUBE_VOLUME
Q AS ENTH_FLOW
TF AS TEMPERATURE
T_EXX AS TEMP_DIFF
WB AS TUBE_LENGTH
DP_DEM AS DELTA_P
FC AS FLOWRATE
FI_IN AS VAP_FLOW
YI_MOLE AS MOLE_FRACTION
P_MIN, P_ORIF AS PRESSURE
DP_ORIF AS DELTA_P
PB_SAT, PD_SAT AS PRESSURE
VD AS VAP_FLOW

STREAM
INPUT 1 F_IN, TF_IN, CF_IN, HF_IN
INPUT 2 D_IN, TD_IN, HD_IN
```

```
INPUT 3 B_IN, TB_IN, PB_IN, CB_IN, HB_IN
INPUT 4 M_IN, TM_IN, CM_IN, HM_IN
OUTPUT 1 F_OUT, TF _OUT, CF_OUT, HF_OUT
OUTPUT 2 D_OUT, TD_OUT, HD_OUT
OUTPUT 3 B_OUT, TB_OUT, PB_OUT, CB_OUT, HB_OUT
OUTPUT 4 R_OUT, TR_OUT, CR_OUT, HR_OUT
INPUT 5 V_IN, TV_IN, PV_IN, YI_IN, HV_IN
OUTPUT 5 V_OUT, TV_OUT, PV_OUT, YI_OUT, HV_OUT


CONNECTION 1 LEVEL
CONNECTION 2 DISTILLATE
CONNECTION 3 FLOW
CONNECTION 4 LEVEL_D
CONNECTION 5 REC_FLOW


EQUATION
```

#Brine Balance

```
#Mass balance of flash chamber
$MB = TON_TO_KG*(B_IN+M_IN -B_OUT-R_OUT -VB);

#Concentration Balance Flash chamber
$MB*CB+$CB*MB
= TON_TO_KG*(B_IN*CB_IN+M_IN*CM_IN -(B_OUT+R_OUT)* CB);

#Enthalpy balance flash chamber
$MB*HB+$HB*MB = TON_TO_KG*(B_IN*HB_IN+M_IN*HM_IN -VB*HV-(B_
OUT+R_OUT)*HB);

#Volume level relationship
MB = RHO_B*AREA_B* LEVEL;

#Assume perfect mixing
HB = HB_OUT;
CB = CB_OUT;
HR_OUT = HB_OUT;

HVB = HVBS+0.48* (TB_OUT-T_SAT);

#Non-equilibrium allowance
TB_OUT = T_SAT+T_EXX;

#Pressure equality
PB_OUT = PB_IN+PAS_TO_BAR*RHO_B*G*LEVEL;
#Partial pressure of inerts
PB_SAT = PB_IN*(1-YI_MOLE);

#Cooling tubes
```

```
#Mass Balance tube side
F_IN = F_OUT = FLOW;

#Concentration Balance tube side
CF_IN = CF_OUT;

#Enthalpy balance tube side
MW*$HF = TON_TO_KG*F_IN*(HF_IN-HF)+Q;

#Heat transfer
Q = UO*AREA_HX*(TD_OUT-T_AVE);
T_AVE = (TF_IN+TF_OUT)/2;

#Assume perfect mixing
HF = HF_OUT;

#Tube holdup
MW = RHO_F*VOL_TUBE*N_TUBE;
VOL_TUBE - = TUBE_AREA*TUBE_LENGTH;

#OHTC in hours
UO_HR = UO*60;

#Assume mixing on tube side
TF = TF_OUT;

#Width of brine chamber is same as tube length
WB = TUBE_LENGTH;

#Vapour space

#Rate of inerts generation
FI_IN = INERTS_RATIO*B_IN;

#Inerts balance
$MI = TON_TO_KG*(V_IN*YI_IN-V_OUT*YI_OUT+FI_IN);

#Mass Balance
$MV = TON_TO_KG*(VB+VD+V_IN-V_OUT+FI_IN-FC);

#Enthalpy Balance
(VB+FI_IN) *HVB+VD*HVD+V_IN*HV_IN = V_OUT*HVV+FC*HVV;

#Volume constraint
V = MV/RHO_V+MD/RHO_D+MB/RHO_B+VOL_TUBE*N_TUBE;

#Mass and mole fraction of inerts
YI_OUT = MI/MV;
YI_MOLE = (MI/MW_INERTS)/((MV-MI)/MW_H2O+MI/MW_INERTS);
```

```
#Flow of vapour extracted
P_MIN = CRIT*PV;
if PV_OUT < P_MIN then
 P_ORIF = P_MIN
else
 P_ORIF = PV_OUT
endif;
DP_ORIF = PV-P_ORIF;
V_OUT = K_ORIF*SQRT (RHO_V*DP_ORIF);

#Vapour streams
PV_IN = PV;

HVV = HVD+0.48*(TV_OUT-TD_OUT);

HV_OUT = HVV;
PV_SAT = PV*(1-YI_MOLE);

#Demister losses
DP_DEM = K_DEM/RHO_V*VB*VB;

#Volume of brine chamber is area*height
V = AREA_B*HEIGHT;

#Condensation rate (approximately-this must not be negative!)
FC = Q/(HV_OUT-HD)/TON_TO_KG;

#Pressure equality
PV = PB_IN-DP_DEM;

#Distillate Balances

#Mass balance
$MD = TON_TO_KG*(D_IN-D_OUT-VD+FC);

#Enthalpy balance
D_IN*HD_IN+FC*HD = D_OUT*HD+VD*HVD;

#Volume/level relationship
MD = RHO_D*AREA_D*LEVEL_D;

#Assume perfect mixing
HD = HD_OUT;

#External connections
D_OUT = DISTILLATE;
REC_FLOW = R_OUT;

#Recycle stream
TR_OUT = TB_OUT;
CR_OUT = CB_OUT;
```

```
#Partial pressure of inerts
PD_SAT = PV*(1-YI_MOLE);

#Property procedures

#Vapour density (neglect inerts)
RHO_V = (MW_H2O*PV)/(R_GAS*(TV_OUT+C_TO_K));

PROCEDURE
(T_EXX) TEXX (stage_no, TB_IN, TB_OUT)
(PB_SAT) PVAP (T_SAT; CB_OUT)
(PD_SAT) PVAP (TD_OUT, C_ZERO)
(HB) BRH (CB-OUT, TB_OUT)
(HVD) STMH (TD_OUT)
(HVBS) STMH (T_SAT)
(RHO_B) BRINRO (CB_OUT, TB_OUT)
(HF_OUT) BRH (CF_OUT, TF_OUT)
(RHO_F) BRINRO (CF_OUT, TF_OUT)
(HD) WATERH (TD_OUT)
(RHO_D) WATERO TD_OUT)
(UO) HTCRIG F_OUT, T_AVE, CF_IN, FC, TD_OUT,
 FOULING, TUBE_ID, TUBE_LENGTH,
 TUBE_THICKNESS, N_TUBE)
(HVB) VAPH (TB_OUT, PB_OUT)
(HVV) VAPH (TV_OUT, PV)
(PV_SAT) PVAP (TV_SAT, C_ZERO)

MODEL ORIFICE
SET
G = 9.81E-S,
ton_to_kg = 1000.0,
bar_to_pas = 1E5;
ton_per_min_to_kg_per_sec = 16.667

TYPE
#stream variables
B_IN, B_OUT;AS FLOWRATE
TB_IN, TB_OUT;AS TEMPERATURE
CB_IN, CB_OUT;AS CONCENTRATION
HB_IN, HB_OUT;AS ENTHALPY
PB_IN, PB_OUT;AS PRESSURE

#connection
LEVEL AS LEVEL#level in upstream stage, m

#set variables
*ho AS LENGTH#height of orifice, m
*wo AS LENGTH#width of orifice, m
*wb AS LENGTH#width of brine channel, m
```

```
#internal variables
RHO_B AS DENSITY
DELTA_P AS DELTA_P
Ao AS AREA#area of orifice, m2
*Cd AS POSITIVE
Cc AS POSITIVE
DP_FACT AS NOTYPE
VC_ht AS NOTYPE
VC_crit AS NOTYPE

STREAM
INPUT B_IN, TB_IN, PB_IN, CB_IN, HB_IN
OUTPUT B_OUT, TB_OUT,PB_OUT, CB_OUT, HB_OUT

CONNECTION 1 LEVEL

EQUATION

#Input/output equalities

B_IN = B_OUT;
TB_IN = TB_OUT;
HB_IN = HB_OUT;
CB_IN = CB_OUT;

#Area of orifice
If LEVEL>ho then
Ao = wo*ho
Else
Ao = wo*LEVEL
Endif;

#Orifice pressure drop (this might go negative-let it-for the
sake of numerical stability)
DELTA_P = PB_IN-PB_OUT+RHO_B*G*(LEVEL-Cc*ho);

#Orifice flow equation (allow negative flow)
DP_FACT = 2*RHO_B*DELTA_P*bar_to_pas;
B_OUT*ABS(B_OUT) = Ao*Ao*Cd*Cd*DP_FACT
/(ton_per_min_to_kg_per_sec*ton_per_min_to_kg_per_sec);

#Vena contracta height vs. critical height
VC_ht = ho*Cc;
VC_crit = (B_OUT*1000.0/60.0/Wo/0.75/RHO_B)'(2.0/3.0)*9.81'-
1.0/3.0);

PROCEDURE
(RHO_B)BRINRO(CB_OUT, TB_OUT)
(Cc)ORIFCD (Wo, ho, Wb, LEVEL, PB_IN, PB_OUT, RHO_B)
#(Cd, Cc)ORFICD(Wo, ho, Wb, LEVEL, PB_IN, PB_OUT, RHO_B)
```

MODEL DL_VALVE

```
TYPE
D_IN, D_OUT AS DIST_FLOW
TD_IN, TD_OUT AS TEMPERATURE
HD_IN, HD_OUT AS ENTHALPY
I_in AS control_signal
#CV AS NOTYPE
FLOW AS DIST_FLOW


STREAM
INPUT D_IN, TD_IN, HD_IN
OUTPUT D_OUT, TD_OUT, HD_OUT


CONNECTION 1 I_in
CONNECTION 200 FLOW


EQUATION

D_IN = D_OUT;
TD_IN = TD_OUT;
HD_IN = HD_OUT;
FLOW = D_OUT;


#Control Valve Characteristics
#D_OUT = CV*I_in;


#Control Valve Characteristics Eq.%
#50.745 t/min = Full Opening Flow
#4.414 = Effective Rangeability
D_OUT = 50.745*(4.414)'(I_in-1);
```

MODEL ST_VALVE
```
SET
Crit_Fact = 0.5
TYPE
S_IN, S_OUT AS FLOWRATE
TS_IN, TS_OUT AS TEMPERATURE
HS_IN, HS_OUT AS VAP_ENTH
PS_IN, PS_OUT AS PRESSURE
I_in AS control_signal
P_lim, P_down AS PRESSURE
DP AS DELTA_P
CV AS NOTYPE
I_loc AS control_signal
SIG_DEL AS NOTYPE
```

```
STREAM
INPUT S_IN, TS_IN, PS_IN, HS_IN
OUTPUT S_OUT, TS_OUT, PS_OUT, HS_OUT
CONNECTION 1 I_in

EQUATION
#general equalities
S_IN = S_OUT;
TS_IN = TS_OUT;
HS_IN = HS_OUT;
#Sonic flow limit
P_1im = Crit_Fact*PS_in;

#Check for sonic flow conditions
IF PS_out>P_lim THEN
P_down = PS_out
ELSE
P_down = P_lim
ENDIF;

#operating pressure difference
DP = PS_in-P_down;
#signal delay
I_Loc = DELAY I_in BY SIG_DEL;
#flow equation.
#(S_OUT*S_OUT)/(CV*CV) = I_Ioc*I_loc*DP;
(S_OUT*S_OUT)/(CV*CV) = 10'(2*(I_loc-1))*DP;
```

MODEL STEAMFEED

```
TYPE
S_OUT AS FLOWRATE
TS_OUT AS TEMPERATURE
PS_OUT AS PRESSURE
HS_OUT AS VAP_ENTH

STREAM
OUTPUT S_OUT, TS_OUT, PS_OUT, HS_OUT

PROCEDURE
(HS_OUT) STMH (TS_OUT)
(HS_OUT) VAPH (TS_OUT, PS_OUT)
```

MODEL DOSING_PUMP

```
TYPE
#connection:
I_in AS CONTROL_SIGNAL
I_out AS CONTROL_SIGNAL
FLOW AS DOSING_FLOW
#CAPACITY ASDOSING_FLOW
```

```
STREAM
CONNECTION 100 I_in#controller signal output to pump
CONNECTION 200 l_out#dosing flow


EQUATION
FLOW = I_in*CAPACITY;
I_out = FLOW;
```

MODEL BL_VALVE

```
TYPE
B_IN, B_OUT AS FLOWRATE
TB_IN, TB_OUT AS TEMPERATURE
CB_IN, CB_OUT AS CONCENTRATION
HB_IN, HB_OUT AS ENTHALPY
PB_IN, PB_OUT AS PRESSURE
I_in AS control_signal
#CV AS NOTYPE
#CVMAX AS NOTYPE
DELP AS PRESSURE


STREAM
INPUT B_IN, TB_IN, PB_IN, CB_IN, HB_IN
OUTPUT B_OUT, TB_OUT, PB_OUT, CB_OUT, HB_OUT


CONNECTION 1 I_in


EQUATION
B_IN = B_OUT;
TB_IN = TB_OUT;
HB_IN = HB_OUT;
CB_IN = CB_OUT;


#PRESSURE EQUATION
PB_OUT = PB_IN-DELP;


#Control Valve Characteristics
#CV = CVMAX*I_IN;
#B_OUT*B_OUT = CV*CV*DELP;
#B_OUT = I_in*CV;


#Valve installed characteristics with EQ.%
#6.9388 = Effective Rangability
#120.2232 t/min = Full openning flow

B_OUT = 120.2232*(6.9388)'(I_in-1);
```

MODEL B_SPLITTER

```
TYPE
FLOW_IN, FLOW_OUT1, FLOW_OUT2 AS FLOWRATE
T_IN, T_OUT1, T_OUT2 AS TEMPERATURE
C_IN, C_OUT1, C_OUT2 AS CONCENTRATION
H_IN, H_OUT1, H_OUT2 AS ENTHALPY
P_IN, P_OUT AS PRESSURE

#Internal variables
*RATIO AS RATIO

STREAM
INPUT 1 FLOW_IN, T_IN, P_IN, C_IN, H_IN
OUTPUT 1 FLOW_OUT1, T_OUT1, P_OUT, C_OUT1, H_OUT1
OUTPUT 2 FLOW_OUT2, T_OUT2, C_OUT2, H_OUT2

EQUATION
#Material Balance
FLOW_IN = FlOW_OUT1+FLOW_OUT2;
FLOW_OUT1 = RATIO*FLOW_IN;

#Concentration Equality
C_IN = C_OUT1 = C_OUT2;

#Temperature Equality
T_IN = T_OUT1 = T_OUT2;

#Enthalpy Equality
H_IN = H_OUT1 = H_OUT2;
#Pressure Equality
P_IN = P_OUT;
```

MODEL CSPLIT

```
TYPE
C_IN, C_OUT1, C_OUT2 AS FLOWRATE
T_IN, T_OUT1, T_OUT2 AS TEMPERATURE
P_IN, P_OUT1, P_OUT2 AS PRESSURE
H_IN, H_OUT1, H_OUT2 AS ENTHALPY

STREAM
INPUT 1 C_IN, T_IN, P_IN, H_IN
OUTPUT 1 C_OUT1, T_OUT1, P_OUT1, H_OUT1
OUTPUT 2 C_OUT2, T_OUT2, P_OUT2, H_OUT2

EQUATION
C_IN = C_OUT1+C_OUT2;
T_IN = T_OUT1 = T_OUT2;
```

```
P_IN = P_OUT1 = P_OUT2;
H_IN = H_OUT1 = H_OUT2;
```

MODEL COND_VALVE

```
TYPE
C_IN, C_OUT AS FLOWRATE
T_IN, T_OUT AS TEMPERATURE
P_IN, P_OUT AS PRESSURE
H_IN, H_OUT AS ENTHALPY
I_in AS CONTROL_SIGNAL
*CV AS FLOWRATE
FLOW AS FLOWRATE

STREAM
INPUT C_IN, T_IN, P_IN, H_IN
OUTPUT C_OUT, T_OUT, P_OUT, H_OUT

CONNECTION 1 I_in
CONNECTION 200 FLOW

EQUATION
C_IN = C_OUT;
T_IN = T_OUT;
H_IN = H_OUT;
FLOW = C_OUT;

#Control Valve-Characteristic
C_OUT = CV*I_in;
```

MODEL MIX

```
TYPE
Flow_IN1, Flow_IN2, Flow_OUT AS FLOWRATE
TF_IN1, TF_IN2, TF_OUT, TEMP AS TEMPERATURE
CF_IN1, Cf_IN2, CF_OUT AS CONCENTRATION
HF_IN1, Hf_In2, HF_OUT AS ENTHALPY

STREAM
INPUT 1 Flow_N1, TF_IN1, CF_IN1, HF_IN1
INPUT 2 Flow_IN2, TF_IN2, CF_IN2, HF_IN2
OUTPUT 1 Flow_OUT, TF_OUT, CF_OUT, HF_OUT

CONNECTION 1 TEMP

EQUATION
TEMP = TF_OUT;
#Mass Balance
Flow_OUT = Flow_IN1+Flow_IN2;
```

```
#Energy Balance
Flow_OUT* HF_OUT = Flow_IN1*HF_IN1+Flow_IN2*HF_IN2;
#Salt balance
Flow_OUT*CF_OUT = Flow_IN1*CF_1N1+Flow_IN2*CF_IN2;

PROCEDURE
(HF_OUT) BRH (CF_OUT,TF_OUT)
```

MODEL MM_SPLITTER

```
TYPE
FLOW_IN, FLOW_OUT1, FLOW_OUT2, TEMP1 AS FLOWRATE
T_IN, T_OUT1, T_OUT2 AS TEMPERATURE
C_IN, C_OUT1, C_OUT2 AS CONCENTRATION
H_IN, H_OUT1, H_OUT2 AS ENTHALPY
#Inte:rnaT variaEles
*RATIO AS RATIO

STREAM
INPUT 1 FLOW_IN, T_IN, C_IN, H_IN
OUTPUT 1 FLOW_OUT1, T_OUT1, C_OUT1, H_OUT1
OUTPUT 2 FLOW_OUT2, T_OUT2, C_OUT2, H_OUI2

CONNECTION 1 TEMP1

EQUATION
TEMP1 = FLOW_IN;
#Material Balance
FLOW_IN = Flow_OUT1+FLOW_OUT2;
FLOW_OUT1 = RATIO*FLOW_IN;

#Concentration Equality
C_IN = C_OUT1 = C_OUT2;
#Temperature Equality
T_IN = T_OUT1 = T_OUT2;
#Enthalpy-Equality
H_IN = H_OUT1 = H_OUT2;
```

MODEL M_SPLITTER

```
TYPE
FLOW_IN, FLOW_OUT1, FLOW_OUT2 AS FLOWRATE
T_IN;T_OUT1, T_OUT2 AS TEMPERATURE
C_IN, C_OUT1, C_OUT2 AS CONCENTRATION
H_IN, H_OUT1, H_OUT2 AS ENTHALPY
#Internal variables
*RATIO AS RATIO

STREAM
INPUT 1 FLOW_IN, T_IN, C_IN, H_IN
OUIPUT 1 FLOW_OUT1, T_OUT1, C_OUT1, H_OUT1
OUTPUT 2 FLOW_OUT2, T_OUT2, C_OUT2, H_OUT2
```

```
EQUATION
#Material Balance
FLOW_IN = Flow_OUT1+FLOW_OUT2;
FLOW_OUT1 = RATIO*FLOW_IN;

#Concentration Equality
C_IN = C_OUT1 = C_OUT2;

#Temperature Equality
T_IN = T_OUT1 = T_OUT2;

#Enthalpy Equality
H_IN = H_OUT1 = H_OUT2;
```

MODEL LAG

```
{Model Version 1.0A
SPEEDUP Model Library Copyright (c) 1991, AspenTech UK Ltd
This copyright statement must not be deleted and must be
included in any modification or adaptation of this Model.}

HELP
First order lag

CONNECTION 100:input signal
CONNECTION 200:output signal

Preferred Sets:gain, Tau
$ENDHELP

TYPE
#connection:
I_in AS control_signal
I_out AS control_signal
#internal:
I AS control_signal#lag output
*Tau AS time#process time constant
*gain AS notype #static gain supplied to signal

STREAM
CONNECTION 100 I_in #input signal
CONNECTION 200 I_out #output signal

EQUATION
#First Order Lag
I+Tau*$I = gain*I_in;
I = I_out;
```

MODEL MFEED

```
TYPE
F_OUT AS FLOWRATE
```

```
TF_OUT AS TEMPERATURE
CF_OUT AS CONCENTRATION
HF_OUT AS ENTHALPY

STREAM
OUTPUT F_OUT, TF_OUT, CF_OUT, HF_OUT

PROCEDURE
(HF_OUT) BRH (CF_OUT, TF_OUT)
```

MODEL PI_CONT

```
{Model Version 1.0A
SPEEDUP Model Library Copyright (c) 1991, AspenTech UK Ltd
This copyright statement must not be deleted and must be
included in any modification or adaptation of this Model.}

HELP
Proportional integral controller
CONNECTION 100:process measurement
CONNECTION 101:setpoint
CONNECTION 200:controller action

Preferred Sets:bias, max, min, pband, reset, SP

Parameter:clip, normal
$ENDHELP

SET
clip = *1,#output clipped
normal = *0 #output unscaled/unnormalised

TYPE
#Connection:
I_in AS.control_signal
*SP AS.control_signal #setpoint
I_out AS.control_signal
#Internal:
*span AS Notype #span of measured value
*bias AS Notype #ss control value
Cont AS notype #
error AS notype #setpnt & variable error
*pband AS notype #proportional band
I_error AS notype #integral of error
*min AS Notype #min. scaled output value
*max AS notype #max. scaled output value
*reset AS notype #integral time (seconds)
value AS notype #calculated value
*action AS notype
```

```
STREAM
CONNECTION 100 I_in #process measurement
CONNECTION 101 SP #setpoint
CONNECTION 200 I_out #controller output

EQUATION
error = (SP-I_in)/span;
$I_error = error;
value = bias+action*100/pband*(error+I_error/reset*60);

#Clip if required
IF clip = 1 THEN
IF value>max THEN
cont = max
ELSE IF value<min THEN
cont = min
ELSE
cont = value
ENDIF
ENDIF
ELSE
cont = value

ENDIF;
#Scale if required
#IF normal>0 THEN
#I_out = normal*(cont-min)/(max-min)
#ELSE
#I_out = cont
#ENDIF;

I_out = cont;
```

MODEL PI_RCONT

```
{Model Version 1.0A
SPEEDUP Model Library Copyright (c) 1991, AspenTech UK Ltd
This copyright statement must not be deleted and must be
included in any modification or adaptation of this Model.}

HELP
Proportional integral controller
CONNECTION 100: primary measurement
CONNECTION 101: reference measurement
CONNECTION 200: controller action
CONNECTION 201: chained measurement

Preferred Sets: ratio bias, max, min, pband, reset
```

```
Parameter: clip, normal
$ENDHELP

SET
clip = *0,#output unclipped
normal = *0 #output unscaled/unnormalised

TYPE
# Connection:
I_in AS control_signal
Ref_in AS control_signal
I_out AS control_signal
I_chain AS control_signal

#Internal:
*span AS notype #span of I_in
Ratio AS notype # set ratio
SP AS control_signal # setpoint
*bias AS notype #ss control value
cont AS notype #
error AS notype #setpnt & variable error
*pband AS notype #controller band
I_error AS notype #integral of error
*min AS notype #min. scaled output value
*max AS notype #max. scaled output value
*reset AS notype #integral time (seconds)
value AS notype #calculated value
*action as notype

STREAM
CONNECTION 100 I_in #process measurement
CONNECTION 101 Ref_in #reference measurement
CONNECTION 200 I_out #controller output
CONNECTION 201 Ref_chain #ref. meas. output
CONNECTION 202 I_chain #proc. meas. output

EQUATION
SP = Ratio*Ref_in;
error = (SP-I_in)/span;
$I_error = error;
value = bias+action*100/pband*(error+I_error/reset*60);
Ref_chain = Ref_in;
I_chain = I_in;

#Clip if required
IF clip = 1 THEN
IF value>max THEN
cont = max
ELSE IF value < min THEN
```

```
cont = min
ELSE
cont = value
ENDIF
ENDIF
ELSE
cont = value
ENDIF;

#Scale if required
#IF normal>0 THEN
#I_out = normal*(cont-min)/(max-min)
#ELSE
#I_out = cont
#ENDIF;

I_out = cont;
```

MODEL REJ_VALVE

```
TYPE
F_IN, F_OUT AS FLOWRATE
TF_IN, TF_OUT AS TEMPERATURE
CF_IN, CF_OUT AS CONCENTRATION
HF_IN, HF_OUT AS ENTHALPY
I_in AS control_signal
#CV AS NO TYPE
FLOW AS FLOWRATE

STREAM
INPUT F_IN, TF_IN, CF_IN, HF_IN
OUTPUT F_OUT, TF_OUT, CF_OUT, HF_OUT

CONNECTION 1 I_in
CONNECTION 200 FLOW

EQUATION
F_IN = F_OUT;
TF_IN = TF_OUT;
HF_IN = HF_OUT;
CF_IN = CF_OUT;
FLOW = F_OUT;

#Control Valve Characteristics
#F_OUT = CV*I_in;

#Control Valve Characteristics Eq.%
#225.34 t/min = Full Open Flow
#2.4730 = Effective Rangeability
F_OUT = 225.34*(5.7345)'(I_in-1);
```

```
MODEL RE1_VALVE

TYPE
B_IN, B_OUT AS FLOWRATE
TB_IN, TB_OUT AS TEMPERATURE
CB_IN, CB_OUT AS CONCENTRATION
HB_IN, HB_OUT AS ENTHALPY
I_in AS control_signal
#CV AS NOTYPE

STREAM
INPUT B_IN, TB_IN, CB_IN, HB_IN
OUTPUT B_OUT, TB_OUT, CB_OUT, HB_OUT

CONNECTION 1 I_in

EQUATION
B_IN = B_OUT;
TB_IN = TB_OUT;
HB_IN = HB_OUT;
CB_IN = CB_OUT;

#Control Valve Characteristics
#B_OUT = CV*I_in;

#Control Valve Characteristics Eq.%
#983.3 833 t/min = Full Valve Open
#9.4977 = Effective Rangeability

B_OUT = 983.3833*(9.4977)'(I_in-1);

MODEL RE2_VALVE

TYPE
F_IN, F_OUT AS FLOWRATE
TF_IN, TF_OUT AS TEMPERATURE
CF_IN, CF_OUT AS CONCENTRATION
HF_IN, HF_OUT AS ENTHALPY
I_in AS control_signal
CV AS NOTYPE
FLOW AS FLOWRATE

STREAM
INPUT F_IN, TF_IN, CF_IN, HF_IN
OUTPUT F_OUT, TF_OUT, CF_OUT, HF_OUT

CONNECTION1 I_in
CONNECTION 200 FLOW
```

```
EQUATION
F_IN = F_OUT;
TF_IN = TF_OUT;
HF_IN = HF_OUT;
CF_IN = CF_OUT;
FLOW = F_OUT

#Control Valve Characteristics
F_OUT = CV*I_in;

#Control Valve Characteristics Eq.%
#300.9417 t/min = Full Valve Open Flow
#2.4730 = Effective Rangeability

F_OUT = 300.9417*(2.4730)'(I_in-1);
```

MODEL MKUP_VALVE

```
TYPE
F_IN, F_OUT AS FLOWRATE
TF_IN, TF_OUT AS TEMPERATURE
CF_IN, CF_OUT AS CONCENTRATION
HF_IN, HF_OUT AS ENTHALPY
I_in AS control_signal
#CV AS NOTYPE
FLOW AS FLOWRATE


STREAM
INPUT F_IN, TF_IN, CF_IN, HF_IN
OUTPUT F_OUT, TF_OUT, CF_OUT, HF_OUT

CONNECTION1 I_in
CONNECTION 200 FLOW


EQUATION
F_IN = F_OUT;
TF_IN = TF_OUT;
HF_IN = HF_OUT;
CF_IN = CF_OUT;
FLOW = F_OUT

#Control Valve Characteristics
F_OUT = CV*I_in;


#Control Valve Characteristics Eq.%
#221.4833 t/min = Full Valve Open Flow
#5.5927 = Effective Rangeability
F_OUT = 221.4833*(5.5927)'(I_in-1);
```

MODEL MKUP_VALVE

```
TYPE
F_IN, F_OUT.AS FLOWRATE
TF_IN, TF_OUT AS TEMPERATURE
CF_IN, CF_OUT AS CONCENTRATION
HF_IN, HF_OUT AS ENTHALPY
I_in AS control_signal
CV AS NOTYPE
FLOW AS FLOWRATE

STREAM
INPUT F_IN, TF_IN, CF_IN, HF_IN
OUTPUT F_OUT, TF_OUT, CF_OUT, HF_OUT

CONNECTION1 I_in
CONNECTION 200 FLOW

EQUATION
F_IN = F_OUT;
TF_IN = TF_OUT;
HF_IN = HF_OUT;
CF_IN = CF_OUT;
FLOW = F_OUT

#Control Valve Characteristics
F_OUT = CV*I_in;

#Control Valve Characteristics Eq.%
#221.4833 t/min = Full open flow
#5.5927 = Effective Rangeability

F_OUT-221.4833*(5.5927)'(I_IN-1);
```

MODEL CUL_VALVE

```
TYPE
F_IN, F_OUT AS FLOWRATE
TF_IN, TF_OUT AS TEMERATURE
CF_IN, CF_OUT AS CONCENTRATION
HF_IN, HF_OUT AS ENTHALPY
I_in AS Control signal
CV AS NOTYPE
FLOW AS FLOWRATE

STREAM
INPUT F_IN, TF_IN, CF_IN, HF_IN
OUTPUT F_OUT, TF_OUT, CF_OUT, HF_OUT

CONNECTION 1 I_in
CONNECTION 200 FLOW
```

```
EQUATION
F_IN = F_OUT;
TF_IN = TF_OUT;
HF_IN = HF_OUT;
CF_IN = CF_OUT;
FLOW = F_OUT;

#Control Valve characteristics
F_OUT = CV*I_in

#Control Valve characteristics Eq%
```

FLOWSHEET
```
#VACUUM SYSTEM
OUTPUT 4 OF F01 IS PRODUCT 7 TYPE VAPOUR
```
FEED 1 IS INPUT 4 OF F02 TYPE VAPOUR
```
OUTPUT 4 OF F02 IS PRODUCT 8 TYPE VAPOUR
```
FEED 2 IS INPUT 4 OF F03 TYPE VAPOUR
```
OUTPUT 4 OF F03 IS PRODUCT 9 TYPE VAPOUR
```
FEED 3 IS INPUT 4 OF F04 TYPE VAPOUR
```
OUTPUT 4 OF F04 IS INPUT 4 OF F05 TYPE VAPOUR
OUTPUT 4 OF F05 IS INPUT 4 OF F06 TYPE VAPOUR
OUTPUT 4 OF F06 IS INPUT 4 OF F07 TYPE VAPOUR
OUTPUT 4 OF F07 IS INPUT 4 OF F08 TYPE VAPOUR
OUTPUT 4 OF F08 IS INPUT 4 OF F09 TYPE VAPOUR
OUTPUT 4 OF F09 IS PRODUCT 10 TYPE VAPOUR
```
FEED 4 IS INPUT 4 OF F10 TYPE VAPOUR
```
OUTPUT 4 OF F10 IS INPUT 4 OF F11 TYPE VAPOUR
OUTPUT 4 OF F11 IS INPUT 4 OF F12 TYPE VAPOUR
OUTPUT 4 OF F12 IS INPUT 4 OF F13 TYPE VAPOUR
OUTPUT 4 OF F13 IS INPUT 4 OF F14 TYPE VAPOUR
OUTPUT 4 OF F14 IS INPUT 4 OF F15 TYPE VAPOUR
OUTPUT 4 OF F15 IS INPUT 4 OF F16 TYPE VAPOUR
OUTPUT 4 OF F16 IS INPUT 4 OF F17 TYPE VAPOUR
OUTPUT 4 OF F17 IS INPUT 5 OF F18 TYPE VAPOUR
OUTPUT 5 OF F18 IS PRODUT 11 TYPE VAPOUR
#BRINE HEATER
OUTPUT 1 OF F01 IS INPUT 1 OF BRINE_HEATER TYPE MAINSTREAM
OUTPUT 1 OF BRINE_HEATER IS INPUT 2 OF F01 TYPE MAINSTREAM
OUTPUT 1 OF SFEED IS INPUT 1 OF ST_VALVE TYPE STEAM
OUTPUT 1 OF ST_VALVE IS INPUT 1 OF DESUP TYPE STEAM
OUTPUT 1 OF DESUP IS INPUT 2 OF BRINE_HEATER TYPE STEAM
OUTPUT 2 OF BRINE_HEATER IS INPUT 1 OF CSPLIT TYPE CONDENSATE
OUTPUT 1 OF CSPLIT IS INPUT 1, OF COND_VALVE TYPE CONDENSATE
OUTPUT 1 OF COND_VALVE IS PRODUCT 1 TYPE CONDENSATE
OUTPUT 2 OF CSPLIT IS INPUT 1 OF DSUP_VALVE TYPE CONDENSATE
OUTPUT 1 OF DSUP_VALVE IS INPUT 2 OF DESUP TYPE CONDENSATE
#FLASH
?repeat
```

```
OUTPUT 1 OF F0 (i+1) IS INPUT 1 OF F0?(i) TYPE MAINSTREAM
OUTPUT 2 OF F0? (i) IS INPUT 2 OF F0? (i+1) TYPE DISTILLATE
OUTPUT 3 of F0? (i) IS INPUT OF W0?(i) TYPE BRINE
OUTPUT OF W0?(i) IS INPUT 3 OF F0?(i+1) TYPE BRINE
?with i = <1:8> *

OUTPUT 1 OF F10 IS INPUT 1 OF F09 TYPE MAINSTREAM
OUTPUT 2 OF F09 IS INPUT 2 OF F10 TYPE DISTILLATE
OUTPUT 3 of F09 IS INPUT OF W09 TYPE BRINE
OUTPUT OF W09 IS INPUT 3 OF F10 TYPE BRINE

#FLASH
?repeat
OUTPUT 1 OF F?(i+1) IS INPUT 1 OF F?(i) TYPE MAINSTREAM
OUTPUT 2 OF F? (i) IS INPUT 2 OF F? (i+1) TYPE DISTILLATE
OUTPUT 3 of F?(i) IS INPUT OF W?(i) TYPE BRINE
OUTPUT OF W?(i) IS INPUT 3 OF F?(i+1) TYPE BRINE
?with i = <10:14>

OUTPUT 2 OF F15 IS INPUT 2 OF F16 TYPE DISTILLATE
OUTPUT 3 OF F15 IS INPUT OF W15 TYPE BRINE
OUTPUT OF W15 IS INPUT 3 OF F16 TYPE BRINE

?repeat
OUTPUT 1 OF F?(i+1) IS INPUT 1 OF F?(i) TYPE MAINSTREAM
OUTPUT 2 OF F?(i) IS INPUT 2 OF F? (i+1)TYPE DISTILLATE
OUTPUT 3 of F?(i) IS INPUT OF W?(i) TYPE BRINE
OUTPUT OF W?(i) IS INPUT 3 OF F?(i+1) TYPE BRINE
?with i = <16:17>
#SPLITTER 1

OUTPUT OF FEED1 IS INPUT 1 OF SPLIT1 TYPE MAINSTREAM
OUTPUT 1 OF SPLIT1 IS INPUT 1 OF MIX TYPE MAINSTREAM
OUTPUT 2 OF SPLIT1 IS INPUT OF CUL_VALVE TYPE MAINSTREAM
OUTPUT OF CUL_VALVE IS PRODUCT 5 TYPE MAINSTREAM

#MIX
OUTPUT OF RE2_VALVE IS INPUT 2 OF MIX TYPE MAINSTREAM
OUTPUT OF MIX IS INPUT 1 OF SPLIT3 TYPE MAINSTREAM
#SPLITTER3
OUTPUT 1 OF SPLIT3 IS INPUT 1 OF F18 TYPE MAINSTREAM
OUTPUT 2 OF SPLIT3 IS PRODUCT 6 TYPE MAINSTREAM

#SPLITTER2
OUTPUT 1 OF F16 IS INPUT 1 OF SPLIT2 TYPE MAINSTREAM
OUTPUT 1 OF SPLIT2 IS INPUT 1 OF MKUP_VALVE TYPE MAINSTREAM
OUTPUT 2 OF SPLIT2 IS INPUT 1 OF SPLIT4 TYPE MAINSTREAM

#SEAWATER MAKEUP VALVE
OUTPUT 1 OF MKUP_VALVE IS INPUT 4 OF F18 TYPE MAINSTREAM
CONNECTION 1 OF MKUP_VALVE IS CONNECTION 200 OF MKUP_CONT
```

```
#SEAWATER MAKEUP CONTROLLER
CONNECTION 100 OF MKUP_CONT IS CONNECTION 200 OF MKUP_VALVE
CONNECTION 101 OF MKUP_CONT IS CONNECTION 2 OF F18

#SPLITTER4
OUTPUT 1 OF SPLIT4 IS INPUT OF RE2_VALVE TYPE MAINSTREAM
OUTPUT 2 OF SPLIT4 IS INPUT OF REJ_VALVE TYPE MAINSTREAM
OUTPUT OF REJ_VALVE IS PRODUCT 4 TYPE MAINSTREAM
OUTPUT 2 OF F18 IS INPUT OF DL_VALVE TYPE DISTILLATE
OUTPUT 3 OF F18 IS INPUT OF BL_VALVE TYPE BRINE
OUTPUT 4 OF F18 IS INPUT OF RE1_VALVE TYPE MAINSTREAM

OUTPUT OF RE1_VALVE IS INPUT 1 OF F15 TYPE MAINSTREAM
OUTPUT OF DL_VALVE IS PRODUCT 2 TYPE DISTILLATE
OUTPUT OF BL_VALVE IS PRODUCT 3 TYPE BRINE

#PASSING THE LEVEL INFORMATION TO THE WEIR UNIT

?repeat
CONNECTION 1 OF F0?(i) IS CONNECTION 1 OF W0?(i)
?with i = <1:9>
?repeat
CONNECTION I OF F?(i) IS CONNECTION 1 OF W?(i)
?with i = <10:17>

#RECYCLE FLOW CONTROLLER
CONNECTION 5 OF F18 IS CONNECTION 100 OF RE1_CONT
CONNECTION 200 OF RE1_CONT IS CONNECTION 1 OF RE1_VALVE

#REJECT SEAWATER CONTROLLER
CONNECTION 3 OF F18 IS CONNECTION 100 OF REJ_CONT
CONNECTION 200 OF REJ_CONT IS CONNECTION 1 OF REJ_VALVE

#BRINE LEVEL CONTROLLER
CONNECTION 1 OF F18 IS CONNECTION 100 OF BL_CONT
CONNECTION 200 OF BL_CONT IS CONNECTION 100 OF BL_VALVE_ACT
CONNECTION 200 OF BL_VALVE_ACT IS CONNECTION 1 OF BL_VALVE
#BRINE HEATER TOP TEMPERATURE CONTROLLER
CONNECTION 1 OF BRINE_HEATER IS CONNECTION 100 OF ST_CONT
CONNECTION 200 OF ST_CONT IS CONNECTION 100 OF ST_VALVE_ACT
CONNECTION 200 OF ST_VALVE_ACT IS CONNECTION 1 OF ST_VALVE

#CONNECTION CUL_VALVE CONTROLLER
CONNECTION 1 OF SPLIT1 IS CONNECTION 100 OF CUL_CONT
CONNECTION 200 OF CUL_CONT IS CONNECTION 1 OF CUL_VALVE

#CONNECTION RE2_VALVE CONTROLLER
CONNECTION 1 OF MIX IS CONNECTION 100 OF RE2_CONT
CONNECTION 200 OF RE2_CONT IS CONNECTION 100 OF RE2_VALVE_ACT
CONNECTION 200 RE2_VALVE_ACT IS CONNECTION 1 OF RE2_VALVE
```

```
#DISTILLATE LEVEL CONTROLLER
CONNECTION 4 OF F18 IS CONNECTION 100 OF DL_CONT
CONNECTION 200 OF DL_CONT IS CONNECTION 1 OF DL_VALVE

#CONNECTION 200 OF DL_CONT IS CONNECTION 100 OF DL_VALVE_ACT
#CONNECTION 200 OF DL_VALVE_ACT IS CONNECTION 1 OF DL_VALVE

#LIME/CAUSTIC SODA DOSING CONTROLLER AND PUMP
CONNECTION 100 OF LIME_PUMP IS CONNECTION 200 OF LIME_CONT
CONNECTION 100 OF LIME_CONT IS CONNECTION 200 OF LIME_PUMP
CONNECTION 101 OF LIME_CONT IS CONNECTION 200 OF DL_VALVE

#ANTI-SCALE DOSING CONTROLLER AND PUMP
CONNECTION 100 OF SCALE_PUMP IS CONNECTION 200 OF SCALE_CONT
CONNECTION 100 OF SCALE_CONT IS CONNECTION 200 OF SCALE_PUMP
CONNECTION 101 OF SCALE_CONT IS CONNECTION 202 OF MKUP_CONT

#SODIUM SULPHITE DOSING CONTROLLER AND PUMP
CONNECTION 100 OF SULP_PUMP IS CONNECTION 200 OF SULP_CONT
CONNECTION 100 OF SULP_CONT IS CONNECTION 200 OF SULP_PUMP
CONNECTION 101 OF SULP_CONT IS CONNECTION 201 OF SCALE_CONT

#BRINE HEATER CONDENSATE LEVEL CONTROLLER
CONNECTION 100 OF COND_CONT IS CONNECTION 2 OF BRINE_HEATER
CONNECTION 1 OF COND_VALVE IS CONNECTION 200 OF COND_CONT

#DESUPERHEATER TEMPERATURE CONTROLLER
CONNECTION 1 OF DESUP IS CONNECTION 100 OF DSUP_CONT
CONNECTION 200 OF DSUP_CONT IS CONNECTION 1 OF DSUP_VALVE

****
UNIT BL_CONT IS A PI_CONT
#set
#clip = 1, normal = 0
****
UNIT BL_VALVE IS A BL_VALVE
****
UNIT BL_VALVE_ACT IS A LAG
****
UNIT BRINE_HEATER IS A BRINE_HEATER
****
UNIT COND_CONT IS A PI_CONT
****
UNIT COND_VALVE IS A COND_VALVE
****
UNIT CSPLIT IS A CSPLIT
****
UNIT CUL_CONT IS A PI_CONT
****
```

```
UNIT CUL_VALVE IS A REJ_VALVE
****
UNIT DESUP IS A DESUP
****
UNIT DL_CONT IS A PI_CONT
#set
#clip = 1, normal = 0
****
UNIT DL_VALVE IS A DL_VALVE
****
UNIT DL_VALVE_ACT IS A LAG
****
UNIT DSUP_CONT IS A PI_CONT
SET CLIP = 1
****
UNIT DSUP_VALVE IS A COND_VALVE
****
UNIT F01 IS A FLASH_FIRST
SET
STAGE_NO = 1
****
UNIT F02 IS A FLASH
SET
STAGE_NO = 2
****
UNIT F03 IS A FLASH
SET
STAGE_NO = 3
****
UNIT F04 IS A FLASH
SET
STAGE_NO = 4
****
UNIT F05 IS A FLASH
SET
STAGE_NO = 5
****
UNIT F06 IS A FLASH
SET
STAGE_NO = 6
****
UNIT F07 IS A FLASH
SET
STAGE_NO = 7
****
UNIT F08 IS A FLASH
SET
STAGE_NO = 8
****
```

```
UNIT F09 IS A FLASH
SET
STAGE_NO = 9
****
UNIT F10 IS A FLASH
SET
STAGE_NO = 10
****
UNIT F11 IS A FLASH
SET
STAGE_NO = 11
****
UNIT F12 IS A FLASH
SET
STAGE_NO = 12
****
UNIT F13 IS A FLASH
SET
STAGE_NO = 13
****
UNIT F14 IS A FLASH
SET
STAGE_NO = 14
****
UNIT F15 IS A FLASH
SET
STAGE_NO = 15
****
UNIT F16 IS A FLASH
SET
STAGE_NO = 16
****
UNIT F17 IS A FLASH
SET
STAGE_NO = 17
****
UNIT F18 IS A FLASH LAST
SET
STAGE_NO = 18
****
UNIT FEED1 IS A MFEED
****
UNIT LIME_CONT IS A PI_RCONT
****
UNIT LIME_PUMP IS A DOSING_PUMP
****
UNIT MIX IS A MIX
****
```

```
UNIT MKUP_CONT IS A PI_RCONT
****
UNIT MKUP_VALVE IS A REJ_VALVE
****
UNIT RE1_CONT IS A PI_CONT
#set
#clip = 1, normal = 0
****
UNIT RE1_VALVE IS A RE_VALVE
****
UNIT RE1_VALVE_ACT IS A LAG
****
UNIT RE2_CONT IS A PI_CONT
#.set
#clip = 1, normal = 0
****
UNIT RE2_VALVE IS A REJ_VALVE
****
UNIT RE2_VALVE_ACT IS A LAG
****
UNIT REJ_CONT IS A PI_CONT
set
#clip = 1, normal = 0
clip = 0, normal = 0
****
UNIT REJ_VALVE IS A REJ_VALVE
****
UNIT REJ_VALVE_ACT IS A LAG
****
UNIT SCALE_CONT IS A PI_RCONT
****
UNIT SCALE_PUMP IS A DOSING_PUMP
****
UNIT SFEED IS STEAMFEED
****
UNIT SPLIT1 IS A MM_SPLITTER
****
UNIT SPLIT2 IS A M_SPLITTER
****
UNIT SPLIT3 IS A M_SPLITTER
****
UNIT SPLIT4 IS A M_SPLITTER
****
UNIT STFL_CONT IS A PI_CONT
****
UNIT ST_CONT IS A PI_CONT
#set
#normal = 0, clip = 1
****
```

```
UNIT ST_VALVE IS A ST_VALVE
****
UNIT ST_VALVE_ACT is a LAG
****
UNIT SULP_CONT IS A PI_RCONT
****
UNIT SULP_PUMP IS A DOSING_PUMP
****
UNIT TEST_HEATER IS A BRINE_HEATER
****
UNIT W01 IS AN ORIFICE
****
UNIT W02 IS AN ORIFICE
****
UNIT W03 IS AN ORIFICE
****
UNIT W04 IS AN ORIFICE
****
UNIT W05 IS AN ORIFICE
****
UNIT W06 IS AN ORIFICE
****
UNIT W07 IS AN ORIFICE
****
UNIT W08 IS AN ORIFICE
****
UNIT W09 IS AN ORIFICE
****
UNIT W10 IS AN ORIFICE
****
UNIT W11 IS AN ORIFICE
****
UNIT W12 IS AN ORIFICE
****
UNIT W13 IS AN ORIFICE
****
UNIT W14 IS AN ORIFICE
****
UNIT W15 IS AN ORIFICE
****
UNIT W16 IS AN ORIFICE
****
UNIT W17 IS AN ORIFICE
```

Operation
SET

#For manual control set VALUE to the output you want and unset
SP or RATIO

Within BL_CONT
SP = 0.6
#VALUE = 0.873506
#VALUE = 0.778411
#VALUE = IF T < 5 THEN 0.778411 ELSE 0.85 ENDIF

Within CUL_CONT
SP = 168.33 {190 181.67}
#VALUE = 0
{VALUE = 0#0.864202E-1}

Within DL_CONT
SP 0.88
#VALUE = 0.880563
#VALUE = 0.860255

Within LIME_CONT
RATIO = 30E-3#30 ppm (30E-6 t/t*1000 l/t)
#VALUE = 0.554132
#VALUE = 0.645191

Within MKUP_CONT
#RATIO = 4.5 {4.837 4.5 4.238}
SP = 91.33 {101.33 92.0 110.38 91.33}
{VALUE = IF T < 5 THEN 0.874956 ELSE 0.774956 ENDIF}
#VALUE = 0.874956

Within RE1_CONT
SP = 238.33 {241.67 90 c}
#SP = 220.08 {106 c}
#VALUE = 0.909819
{VALUE IF T < 5 THEN 0.909819 ELSE 0.889819 ENDIF}

#SP = {230.0} {241.83} {238.33}

Within RE2_CONT
SP = 26
#SP = IF T<1 THEN 26 ELSE 24 ENDIF
#VALUE = 0
{VALUE = 0#0.399846}

Within REJ_CONT
SP = 220.0 {241.67} {208.33} {241.8#205.0}
#VALUE 1.015503
{SP = 241.67#(summer)}
#VALUE = 0.463708

Within SCALE_CONT
RATIO 2.0E-3#2 ppm (2E-6 t/t*1000 l/t)
#VALUE = 0.357378
#VALUE = 0.387115

Within ST_CONT
```
#SP = 89
SP = IF T < 1 THEN 89 ELSE 92 ENDIF
#VALUE = 0.637623
#VALUE = 0.72
{VALUE = 0.943130}

{VALUE = IF T < 5 THEN 0.943130 ELSE 0.893130 ENDIF}
```

Within SULP_CONT
```
RATIO 5.0E-3#5 ppm (5E-6 t/t*1000 l/t)
#VALUE 0.446723
#VALUE = 0.483893
```

Within COND_CONT
```
SP 0.15#m
#VALUE = 0.660972
```

Within DSUP_CONT
```
SP = 105 {C}
#VALUE = 0.765401
#VALUE = 0.5
```

WITHIN DSUP_CONT#
```
SPAN = 120#C
PBAND = 50#%
RESET = 60#sec/repeat
ACTION = -1
BIAS = 0
MAX = 1
MIN = 0
```

WITHIN DSUP_VALVE
```
CV = 0.06#t/min = 4 t/h
```

WITHIN DESUP
```
C_ZERO = 0#MUST BE ZERO
```

WITHIN COND_CONT#
```
SPAN = 2.0#m
PBAND = 25#%
RESET = 600   #sec/repeat
ACTION = -1
BIAS = 0
MAX = 1
MIN = 0
```

WITHIN COND_ VALVE
```
CV = 4.2#t/min = 250 t/h
P_OUT = 0.5
```

WITHIN SCALE_PUMP
```
CAPACITY = 0.5
```

WITHIN SCALE_CONT#WW00-F001
SPAN = 0.5# 30 l/h
PBAND = 200#%
RESET = 60#sec/repeat
ACTION = 1
BIAS = 0
MAX = 0.5#l/min
MIN = 0

WITHIN SULP_PUMP
CAPACITY = 1.0

WITHIN SULP_CONT#WJ00-F001
SPAN = 1.0#60 l/h
PBAND = 200#%
RESET = 60#sec/repeat
ACTION = 1
BIAS = 0
MAX = 1#l/min
MIN = 0

WITHIN LIME_PUMP
CAPACITY = 1.0

WITHIN LIME_CONT#WQ00-F001
SPAN = 1.0#60 l/h
PBAND = 200   #%
RESET = 60#sec/repeat
ACTION = 1
BIAS = 0
MAX = 1#l/min
MIN = 0

WITHIN MKUP_VALVE
CV = 121.7

WITHIN MKUP_CONT#WD18-F001
SPAN = 133.33#8000 t/h
PBAND = 48#%
RESET = 22#sec/repeat
ACTION = 1
BIAS = 0
MAX = 1
MIN = 0

WITHIN ST_VALVE
{CV = 3.33#5.642}
CV = 6.6#4.2
SIG_DEL = 0.4#DELAY BEFORE ACTION STARTS
#I_IN = 0.5

```
WITHIN ST_VALV_ACT
GAIN = 1
TAU = 1

WITHIN ST_CONT#WF15-T002
SPAN = 120#C
PBAND = 100{25%}
RESET = 600{32#sec/rpt}
ACTION = 1
BIAS = 0
MAX = 1.5#1
MIN = 0

WITHIN RE1_VALVE
CV = 265.8#483.660
#I_IN = 0.5
{WITHIN REI_VALVE_ACT
GAIN = 1
TAU = 0.1}

WITHIN RE1_CONT#WF12-F001
SPAN = 333.333#t/min = 20,000 t/h
PBAND 60
RESET = 29
ACTION = 1
BIAS = 0
MAX = 1
MIN = 0

WITHIN RE2_VALVE
CV = 96.7#193.33
#I_IN = 0.5

WITHIN RE2_VALVE_ACT
GAIN = 1
TAU = 0.1

WITHIN RE2_CONT#WD10-T020
SPAN = 50#C
PBAND = 20
RESET = 34
ACTION = 1
BIAS = 0
MAX = 1
MIN = 0

WITHIN REJ_VALVE
CV = 150#299.560
#F_OUT = 149.78
#I_IN = 0.5
```

```
{WITHIN REJ_VALVE_ACT
GAIN = 1
TAU = 1}

WITHIN REJ_CONT#ND10-F004
SPAN = 333.333#20,000 t/h
PBAND = 31
RESET = 24
ACTION = 1
BIAS = 0
MAX = 1
MIN = 0

WITHIN BL_CONT#WI18-L001
SPAN = 2#m
PBAND = 67#20
RESET = 10000#37
ACTION = -1
BIAS = 0
MAX = 1
MIN = 0

WITHIN BL_VALVE
CV = 96.7
#CVMAX = 603.757
#I_in = 0.5
#CV = 299.338#145.418
#DELP = 0.059
PB_OUT = 0.0819

WITHIN BL_VALVE_ACT
GAIN = 1
TAU = 0.5

WITHIN DL_CONT#WI18-L004
SPAN = 1#m
PBAND = 40
RESET = 38
ACTION = -1
BIAS = 0
MAX = 1
MIN = 0

WITHIN DL_VALVE
CV = 25#45.184
#I_ in = 0.5

{WITHIN DL_VALVE_ACT
GAIN = 1
TAU = 1#0.5}
```

WITHIN CUL_VALVE
CV = 61.7#110

WITHIN CUL_CONT#WD10-F001
SPAN = 333.333#20,000 t/h
PBAND = 20
ACTION = 1
BIAS = 0
MAX = 1
MIN = 0

WITHIN F01
INERTS_RATIO = 0.2139E-4
K_ORIF = 0.9227E-1
MW_INERTS 43.18
#fi_in = 5.17E-3
#v_out = 32.7E
PV_OUT = 0.1
TUBE_THICKNESS = 1.22#mm
C_ZERO = 0
HEIGHT = 4.27
K_DEM = 2.3E-4
#DP_DEM = 1.08E-3 {11 mmH2O}
DIST_COF = 83.179
AREA_B = 61.885
AREA_HX = 4535.8
AREA_D = 3.6238
N_TUBE = 2860
TUBE_AREA = 6.74716E-04
TUBE_ID = 30.0
FOULING = 0.0002#hr-m2-C/kcal
TUBE_LENGTH = 15.9
#level = 0.2 {0.180 0.305}
#level_d = 0.02
{UO = 41.7518}
{TLOSS = 1.2}
#pvap = 0.61
? repeat
WITHIN F0?(i)
HEIGHT = 4.27
AREA_B = 61.885
AREA_HX = 4535.8
AREA_D = 3.6309
N_TUBE = 2860
TUBE_AREA = 6.74716E-04
TUBE_LENGTH = 15.9
TUBE_ID = 30.0
FOULING = 0.0002#hr-m2-C/kcal
?with i = <2:3>

WITHIN F02
```
V_IN = 0
TV_IN = 30
HV_IN = 500
YI_IN = 0
INERTS_RATIO = 0.4284E-5
K_ORIF = 0.1915E-1
MW_INERTS = 43.18
#fi_in = 1.03E-3
#v_out = 6.097E-3
PV_OUT = 0.1
K_DEM = 2.1E-4
#DP_DEM = 1.08E-3 {11 mmH2O}
TUBE_THICKNESS = 1.22#mm
C_ZERO = 0
DIST_COF = 64.753
#level = 0.500 {0.250 0.310}
#level_d = 0.05
{UO = 41.5462}
{TLOSS = 1.5#1.253}
#pvap = 0.54
```

WITHIN F03
```
V_IN = 0
TV_IN = 30
HV_IN = 500
YI_IN = 0
INERTS_RATIO = 0.9072E-6
K_ORIF = 0.4529E-2
MW_INERTS = 42.11
#fi_in = 0.217E-3
#v_out = 1.288E-3
PV_OUT = 0.1
K_DEM = 1.9E-4
#DP_DEM = 1.08E-3 {11 mmH2O}
TUBE_THICKNESS = 1.22#mm
C_ZERO = 0
DIST_COF = 40.896
#level = 0.250 {0.320#0.313}
#level_d = 0.12
{UO = 41.3163}
(TLOSS = 1.0#1.274}
#pvap = 0.48
? repeat
WITHIN F0?(i)
HEIGHT = 4.27
AREA_B = 61.885
AREA_HX = 4535.8
AREA_D = 3.7312
```

```
N_TUBE = 2860
TUBE_ AREA = 6.74716E-04
TUBE_LENGTH = 15.9
TUBE_ID = 30.0
FOULING = 0.0002#hr-m2-C/kcal
?with i = <4:7>
```

WITHIN F04
```
V_IN = 0
TV_IN = 30
HV_IN = 500
YI_IN = 0
INERTS_RATIO = 0.3277E-6
K_ORIF = 0.2099E-1
MW_INERTS = 37.03
#fi_in = 0.078E-3
#v_out = 2.377E-3
K_DEM = 1.8E-4
#DP_DEM = 1.08E-3 {11 mmH2O}
TUBE_THICKNESS = 1.22#mm
C_ZERO = 0
DIST_COF = 59.730
level = 0.400 {0.340#0.318}
level_d = 0.110
{UO = 42.9156}
{TLOSS = 1.1#1.317}
#pvap = 0.43
```

WITHIN F05
```
INERTS_RATIO = 0.1550E-6
K_ORIF = 0.3554E-1
MW_INERTS = 34.0
#fi_in = 0.0367E-3
#v_out = 3.613E-3
K_DEM = 1.6E-4
#DP_DEM = 1.08E-3 {11 mmH2O}
TUBE_THICKNESS = 1.22#mm
C_ZERO = 0
DIST_COF = 30.339
#level = 0.320 {0.350#0.323}
#level d = 0.270
{UO = 42.6360}
{TLOSS = 1.1#1.359}
#pvap = 0.38
```

WITHIN F06
```
INERTS_RATIO = 0.1795E-6
K_ORIF = 0.5619E-1
MW_INERTS = 32.49
#fi_in = 0.0423E-3
```

```
#v_out = 5.094E-3
K_DEM = 1.5E-4
#DP_DEM = 1.08E-3 {11 mmH2O}
TUBE_THICKNESS = 1.22#mm
C_ZERO = 0
DIST_COF = 44.386
#level = 0.350-{0.330#0.328}
#level_d = 0.220
{UO = 42.3498}
{TLOSS = 1.1#1.402}
#pvap = 0.33
```

WITHIN F07
```
INERTS_RATIO = 0.1961E-6
K_ORIF = 0.8155E-1
MW_INERTS = 31.62
#fi_in = 0.046E-3
#v_out = 6.589E-3
K_DEM = 1.3E-4
#DP_DEM = 1.08E-3 {11 mmH2O}
TUBE_THICKNESS = 1.22#mm
C_ZERO = 0
DIST_COF = 39.644
#level = 0.300 {0.370#0.334}
#level_d = 0.280
{TLOSS = 1.9#1.465}
{UO = 42.053}
#pvap = 0.29
```

WITHIN F08
```
INERTS_RATIO = 0.1971E-6
K_ORIF = 0.1100
MW_INERTS = 31.09
#fi_in = 0.046E-3
#v_out = 7.913E-3
K_DEM = 1.2E-4
#DP_DEM = 1.08E-3 {11 mmH2O}
TUBE_THICKNESS = 1.22#mm
C_ZERO = 0
HEIGHT = 4.27
DIST_COF = 50.189
AREA_B = 61.885
AREA_HX = 4535.8
AREA_D = 3.7312
N_TUBE = 2860
TUBE_AREA = 6.74716E-04
TUBE_LENGTH = 15.9
TUBE_ID = 30.0
FOULING = 0.0002#hr-m2-C/kcal
#level = 0.380 {0.340#0.339}
```

```
#level_d = 0.250
{UO = 41.745}
{TLOSS = 1.2#1.4508}
#pvap = 0.25
```

WITHIN F09
```
PV_OUT = 0.05
INERTS_RATIO = 0.1980E-6
K_ORIF = 0.7169E-1
MW_INERTS = 30.74
#fi_in = 0.046E-3
#v_out = 9.263E-3
K_DEM = 1.1E-4
#DP_DEM = 1.08E-3 {11 mmH2O}
TUBE_THICKNESS = 1.22#mm
C_ZERO = 0
HEIGHT = 4.27
DIST_COF = 63.340
AREA_B = 61.885
AREA_HX = 4535.8
AREA_D = 3.6539
N_TUBE = 2860
TUBE_AREA = 6.74716E-04
TUBE_LENGTH = 15.9
TUBE_ID = 30.0
FOULING = 0.0002#hr-m2-C/kcal
#level = 0.340 {0.320#0.345}
#level_d = 0.220
{UO = 41.436}
{TLOSS = 1.2#1.561}
#pvap = 0.22
```

WITHIN F10
```
V_IN = 0
TV_IN = 30
HV_IN = 500
YI_IN = 0
INERTS_RATIO = 0.1990E-6
K_ORIF = 0.2462E-1
MW_INERTS = 29.0
#fi_in = 0.046E-3
#v_out = 1.40E-3
K_DEM = 1.0E-4
#DP_DEM = 1.08E-3 {11 mmH2O}
TUBE_THICKNESS = 1.22#mm
C_ZERO = 0
HEIGHT = 3.89
DIST_COF = 101.587
AREA_B = 61.885
AREA_HX = 4535.8
```

```
AREA_D = 4.9315
N_TUBE = 2860
TUBE_AREA = 6.74716E-04
TUBE_LENGTH = 15.9
TUBE_ID = 30.0
FOULING = 0.0002#hr-m2-C/kcal
#level = 0.660 {0.40#0.351}
#level_d = 0.150
{UO = 41.3163}
{TLOSS = 1.3#1.635}
#pvap = 0.19
?repeat
```

WITHIN F0?(i)
```
HEIGHT = 3.89
AREA_B = 61.885
AREA_HX = 4535.8
AREA_D = 5.2192
N_TUBE = 2860
TUBE_AREA = 6.74716E-04
TUBE_LENGTH = 15.9
TUBE_ID = 30.0
FOULING = 0.0002#hr-m2-C/kcal
?with i = <11:13>
```

Within F11
```
INERTS_RATIO = 0.1999E-6
K_ORIF = 0.5623E-1
MW_INERTS = 29.0
#fi_in = 0.046E-3
#v_out = 2.818E-3
K_DEM = 1.0E-4
#DP_DEM = 1.08E-3 {11 mmH2O}
TUBE_THICKNESS = 1.22#mm
C_ZERO = 0
DIST_COF = 109.856
#level = 0.680 {0.390#0.356}
#level_d = 0.150
{TLOSS = 1.3#1.688}
{UO = 40.7805}
#pvap = 0.17
```

WITHIN F12
```
INERTS_RATIO = 0.2008E-6
K_ORIF = 0.9316E-1
MW_INERTS = 29.0
#fi_in = 0.046E-3
#v_out = 4.124E-3
K_DEM = 1.0E-4
#DP_DEM = 1.08E-3 {11 mmH2O}
```

```
TUBE_THICKNESS = 1.22#mm
C_ZERO = 0
DIST_COF = 84.071
#level = 0.700 {0.580#0.362}
#level_d = 0.210
{TLOSS = 1.3#1.741}
{UO = 40.4442}
#pvap = 0.15
```

WITHIN F13
```
INERTS_RATIO = 0.2017E-6
K_ORIF = 0.1395
MW_INERTS = 29.0
#fi_in = 0.046E-3
#v_out = 5.432E-3
K_DEM = 1.0E-4
#DP_DEM = 1.08E-3 {11 mmH2O}
TUBE_THICKNESS = 1.22#mm
C_ZERO = 0
DIST_COF = 89.369
#level = 0.680 {0.6#0.367}
#level_d = 0.210
{TLOSS = 1.3#1.826}
{UO = 40.1030}
#pvap = 0.13
?repeat
```

WITHIN F0?(i)
```
HEIGHT = 3.89
AREA_B = 61.885
AREA_HX = 4535.8
AREA_D = 5.2192
N_TUBE = 2860
TUBE_ AREA = 6.74716E-04
TUBE_LENGTH = 15.9
TUBE_ID = 30.0
FOULING = 0.0002#hr-m2-C/kcal
TUBE_LENGTH = 15.9
?with i<14:15>
```

WITHIN F14
```
INERTS_RATIO = 0.2026E-6
K_ORIF = 0.1942
MW_INERTS = 29.0
#fi_in = 0.046E-3
#v_out = 6.641E-3
K_DEM = 1.0E-4
#DP_DEM = 1.08E-3 {11 mmH2O}
TUBE_THICKNESS = 1.22#mm
C_ZERO = 0
```

```
DIST_COF = 89.979
#level = 0.650 {0.6#0.371}
#level_d = 0.220
{TLOSS = 1.4#1.911}
{UO = 39.7638}
#pvap = 0.12
```

WITHIN F15
```
INERTS_RATIO = 0.2035E-6
K_ORIF = 0.2850
MW_INERTS = 29.0
#fi_in = 0.046E-3
#v_out = 7.833E-3
K_DEM = 0.9E-4
#DP_DEM = 1.08E-3 {11 mmH2O}
TUBE_THICKNESS = 1.22#mm
C_ZERO = 0
DIST_COF = 76.897
#level = 0.630 {0.6#0.382}
#level_d = 0.270
{TLOSS = 1.4#1.986}
{UO = 39.4330}
#pvap = 0.1
```

WITHIN F16
```
INERTS_RATIO = 0.2044E-6
K_ORIF = 0.3643
MW_INERTS = 29.0
#fi_in = 0.046E-3
#v_out = 9E-3
K_DEM = 1.3E-4
#DP_DEM = 1.08E-3 {11 mmH2O}
TUBE_THICKNESS = 1.22#mm
C_ZERO = 0
HEIGHT = 3.89
DIST_COF = 73.865
AREA_B = 59.218
AREA_HX = 4306.0
AREA_D = 5.2192
N_TUBE = 2715
TUBE_AREA = 6.74716E-04
TUBE_LENGTH = 15.9
TUBE_ID = 30.0
FOULING = 0.00023#hr-m2-C/kcal
#level = 0.570 {0.570#0.385}
#level_d = 0.290
{UO = 35.7377}
{TLOSS = 1.5#2.25}
#pvap = 0.09
```

WITHIN F17
INERTS_RATIO = 0.2051E-6
K_ORIF = 0.4836
MW_INERTS = 29.0
#fi_in = 0.046E-3
#v_out = 10.17E-3
K_DEM = 1.2E-4
#DP_DEM = 1.08E-3 {11 mmH2O}
TUBE_THICKNESS = 1.22#mm
C_ZERO = 0
HEIGHT = 3.89
DIST_COF = 78.973
AREA_B = 59.218
AREA_HX = 4306.0
AREA_D = 5.2192
N_TUBE = 2715
TUBE_AREA = 6.74716E-04
TUBE_LENGTH = 15.9
TUBE_ID = 30.0
FOULING = 0.00023#hr-m2-C/kcal
#level = 0.540 {0.550#0.389}
#level_d = 0.280
{UO = 35.5228}
{TLOSS = 1.4#2.336}
#pvap = 0.08

WITHIN F18
INERTS_RATIO = 0.2058E-6
K_ORIF = 0.2618
MW_INERTS = 29.0
#fi_in = 0.046E-3
#v_out = 11.33E-3
K_DEM = 0.6E-4
#DP_DEM = 1.08E-3 {11 mmH2O}
TUBE_THICKNESS = 1.22#mm
C_ZERO = 0
HEIGHT = 3.89
AREA_B = 59.218
AREA_HX = 4306.0
AREA_D = 5.4717
N_TUBE = 2715
TUBE_AREA = 6.74716E-04
TUBE_LENGTH = 15.9
TUBE_ID = 30.0
FOULING = 0.00023#hr-m2-C/kcal
#level = 0.588
{UO = 35.2335}
{TLOSS = 1.3#2.27}
#pvap = 0.07

```
WITHIN W01
ho = 0.114#height of orifice, m
wo = 12.682#width of orifice, m
wb = 15.9#width of brine chamber, m
cd = 0.651886
WITHIN W02
ho = 0.117
wo = 12.682
wb = 15.9
cd = 0.558368
WITHIN W03
ho = 0.121
wo = 12.682
wb = 15.9
cd = 0.542194
WITHIN W04
ho = 0.125
wo = 12.682
wb = 15.9
cd = 0.597273
WITHIN W05
ho = 0.129
wo = 12.682
wb = 15.9
cd = 0.631425
WITHIN W06
ho = 0.133
wo = 12.682
wb = 15.9
cd = 0.620609
WITHIN W07
ho = 0.137
wo = 12.682
wb = 15.9
cd = 0.651431
WITHIN W08
ho = 0.142
wo 12.682
wb = 15.9
cd = 0.606820
WITHIN W09
ho = 0.120
wo = 10.74
wb = 15.9
cd = 0.890053
WITHIN W10
ho = 0.215
wo = 12.682
wb = 15.9
cd = 0.34876
```

```
WITHIN W11
ho = 0.220
wo = 12.682
wb = 15.9
cd = 0.341962
WITHIN W12
ho = 0.225
wo = 12.682
wb = 15.9
cd = 0.334366
WITHIN W13
ho = 0.230
wo = 12.682
wb = 15.9
cd = 0.336140
WITHIN W14
ho = 0.235
wo = 12.682
wb = 15.9
cd = 0.340377
WITHIN W15
ho = 0.237
wo = 12.682
wb = 15.9
cd = 0.350947
WITHIN W16
ho = 0.240
wo = 12.682
wb = 15.9
cd = 0.368822
WITHIN W17
ho = 0.240 {0.320}
wo = 12.682
wb = 15.9
cd = 0.382854

WITHIN FEED1
TF_OUT = 22
#TF_OUT-IF T< 5 THEN 22 ELSE 20 ENDIF
CF_OUT = 0.05
#F_OUT = 196.5

WITHIN SFEED
TS_OUT = 130
#TS_OUT = IF T < 5 THEN 105 ELSE 108 ENDIF
PS_OUT = 1.4 {1.7} {1.8} {2.0}
#S_OUT = 2.804

WITHIN SPLIT2
#RATIO = 0.47376
```

WITHIN SPLIT3
#RATIO = 0.9179
FLOW_OUT2 = 18.333

#within split4
#flow_out2 = 149.771

WITHIN BRINE HEATER

{U = 34.6#OHTC}
AHX = 4664#AREA FOR HEAT EXCHANGE
MB = 35500#MASS OF BRINE IN TUBES
V = 94.5#VOLUME OF SHELL PLUS SUMP
ASUMP = 19.7#X-SECTIONAL AREA OF SUMP
RHO_C = 1000#CONDENSATE DENSITY
ZERO_CONC = 0#MUST BE ZERO!
SIG_DEL = 0.4#DELAY TO TEMPERATURE TRANSMITTER
MHX = 101800#EMPTY MASS OF HEAT EXCHANGER
TUBE_LENGTH = 17.33#m
TUBE_ID = 29.31#mm
N_TUBE = 2700#
TUBE_THICKNESS = 1.22#mm
FOULING = 0.00035#m2-K/(kcal/hr)

PRESET

WITHIN RE2_VALVE
F_IN = 52:0:110

WITHIN SPLIT1
RATIO = 0.939:0:1

WITHIN SPLIT4
RATIO = 0.4233098:0:1

10.2 Subroutines in FORTRAN

10.2.1 SUBROUTINE BRINRO (CB, TB, RO) to Compute the Density of Brine

```
SUBROUTINE BRINRO (CB, TB, RO)
C      RO    : DENSITY OF BRINE SOLUTION KG/CU.M
C      TB    : BRINE TEMPERATURE         C
C      CB    : BRINE CONCENTRATION, MASS FRACTION
C
IMPLICIT DOUBLE PRECISION (A-Z)
TB1 = TB*1.8 +32.
RO1 = 62.707172 + 49.364088*CB - 0.43955304E-02*TB1
```

```
&        - 0.032554667*CB*TB1 - 0.46076921E-04*TB1*TB1
&        + 0.63240299E-04*CB*TB1**2
RO = RO1 * 16.0256
RETURN
END
```

10.2.2 SUBROUTINE WATRCP (TB, SD) to Compute
the Water Heat Capacity

```
SUBROUTINE WATRCP (TB, SD)
IMPLICIT DOUBLE PRECISION (A-Z)
C        TB      : TEMPERATURE          c
C        SD      : HEAT CAPACITY,    kcal/kg        C
TB1 = TB * 1.8 + 32.
SD = 1.0011833 - 6.1666652E - 05*TB1+1.3999989E-07*
&        TB1*TB1 + 1.3333336E-09*TB1**3
RETURN
END
```

10.2.3 SUBROUTINE BRINCP (CB, TB, SB)
to Compute the Brine Heat Capacity

```
C+++ Heat Capacity of Brine
SUBROUTINE BRINCP (CB, TB, SB)
IMPLICIT DOUBLE PRECISION (A-Z)
C        TB      : TEMPERATURE          C
C        SB      : HEAT CAPACITY,    kcal/kg        C
CALL WATRCP (TB, SD)
TB1 = TB * 1.8 + 32.
CBP = CB*100.
A = CBP*0.011311
B = CBP*0.00001146
SB = (1.0 - (A-B*TB1)) * SD
RETURN
END
```

10.2.4 SUBROUTINE WATERH (TB, HD) to Compute
the Specific Enthalpy of Water

```
SUBROUTINE WATERH (TB, HD)
IMPLICIT DOUBLE PRECISION (A-Z)
C        HD      : SPECIFIC ENTHALPY OF WATER (KCAL/KG)
C        TB      : THE BOILING TEMPERATURE C
TB1 = TB * 1.8 + 32.
HD1 = -31.92 + 1.0011833*TB1 - 3.0833326E-05*TB1*TB1
&        + 4.666663E-08*TB1**3 + 3.333334E-10*TB1**4
HD = HD1/1.8
RETURN
END
```

10.2.5 SUBROUTINE BRH (C, T, H) to Compute the Specific Enthalpy of Brine

```
Subroutine BRH (C, T, H)
C      T      : temperature
C      C      : concentration
C      H      : enthalpy (Kcal/kg)
IMPLICIT DOUBLE PRECISION (A-Z)
CALL RHO (C, T, RO)
C         convert the unit of s to (g/l)
S = C*RO*1000.
A = 4.185 - 5.381E-3*S+6.260E-6*S**2
B = 3.055E-5 + 2.774E-6*S-4.318E-8*S**2
L = 8.844E-7 + 6.527E-8*S-4.003E-10*S**2
K = 4.1868
H = A*T/K-B*T**2/2/K+L*T**3/3/K
RETURN
END
```

10.2.6 SUBROUTINE STMH (T, H) to Compute the Saturated Steam Specific Enthalpy

```
SUBROUTINE STMH (T, H)
C      T      : temperature
C      H      : Saturated steam enthalpy (Kcal/kg)
IMPLICIT DOUBLE PRECISION (A-Z)
H = (2499.15+1.955*T-1.927E-3*T**2)/4.1868
RETURN
END
```

10.2.7 SUBROUTINE RHO (C, T, RO) to Calculate the Density of Brine

```
SUBROUTINE RHO (C, T, RO)
C      T      ; Temperature
C      C      : Concentration c
C      RO     : The density (g/cm3 or kg/l)
IMPLICIT DOUBLE PRECISION (A-Z)
S = C*1000.
Y = (2.*T-200.)/160.
SIGMA = (2.*S-150.)/150
A0 = 2.016110 + 0.115313*SIGMA+
&      0.000326*(2.*SIGMA**2-1.)
A1 = -0.05410 + 0.001571*SIGMA-
&      0.000423*(2.*SIGMA**2-1.)
A2 = -0.006124 + 0.001740*SIGMA-
&      0.000009*(2.*SIGMA**2-1.)
A3 = 0.000346 + 0.000087*SIGMA-
&      0.000053*(2.*SIGMA**2-1.)
RO =.5*A0+A1*Y+A2*(2.*Y**2-1) +A3*(4*Y**3-3.*Y)
RETURN
END
```

10.2.8 SUBROUTINE EXX (TBIN, TBOUT, TD, WD, HT, W, EX) to Calculate the Nonequilibrium Allowance

```
SUBROUTINE EXX (TBIN, TBOUT, TD, WD, HT, W, EX)
IMPLICIT DOUBLE PRECISION (A-Z)
C       TBIN,TBOUT,TD : TEMPERATURE        C
C       WD            : STAGE WIDTH,       M
C       HT            : BRINE POOL LEVEL, M
C       EX            : NON-EQUILIBRATION ALLOWANCE,C
C       W             : COOLING BRINE FLOWRATE,  t/min
C       TYPE *, 'TBIN, TBOUT, TD, WD, HT, W, EX'
C       TYPE *, TBIN, TBOUT, TD, WD, HT, W, EX

W1      = W * 2200.0 * 60.
IF (HT.LT.0.3) HT = 0.3
HT1 = HT * 12.0/0.3048
DTB = (TBIN - TBOUT) *1.8
OMEGA = W1/(WD/0.3048)
A1 = HT1**1.1
IF (DTB.LE.0.1)DTB = 0.1
A2 = DTB**(-0.25)
A3 = (OMEGA*1.0E-03)**.5
TD1 = TD * 1.8 + 32.
IF(TD1.LT.40)TD1 = 40.0
A4 = TD1**(-2.5)
EX = (352.*A1*A2*A3*A4)/1.8
RETURN
END
```

10.2.9 SUBROUTINE TLOSS (TD, DELT) to Compute the Temperature Loss across the Demister and Condenser Tubes

```
c ++++ Temperature loss across demister & condenser tubes F
SUBROUTINE TLOSS (TD, DELT)
IMPLICIT DOUBLE PRECISION (A-Z)
C       TD            : TEMPERATURE C
C       DELT          : LOSS        C
TD1 = TD *1.8 + 32.
DELT = (EXP(1.885- 0.02063*TD1))/1.8
RETURN
END
```

10.2.10 SUBROUTINE PVAP (T, C, PV) to Compute the Seawater Vapor Pressure

```
SUBROUTINE PVAP (T, C, PV)
C       T        : Temperature C
C       C        : Concentration kg/kg
C       PV       : Seawater Vapor pressure Bar
IMPLICIT DOUBLE PRECISION (A-Z)
```

```
Tk = 647.25
Pk = 220.93
S = C*1000.
Ta = T + 273.15
L = (Tk/Ta)*
&((-7.8889166*(1-Ta/Tk))+(2.5514255*((1-
&Ta/Tk)**1.5))
&+(-6.7161690 *((1 - Ta/Tk)**2)) + (33.239495 *((1 -
&Ta/Tk)**2.5))
&+(-105.38479 *((1 - Ta/Tk)**3)) + (174.35319 *((1 -
&Ta/Tk)**3.5))
&+(-148.39348 *((1 - Ta/Tk)**4)) + (48.631602 *((1 -
&Ta/Tk)**4.5)))
Pw = Pk * 2.7183**L
Pv = Pw * (1 - 0.000537*S)
RETURN
END
```

10.2.11 SUBROUTINE WATERO (T, RO) to Compute the Density of Water

```
SUBROUTINE WATERO (T, RO)
C      T: TEMPERATURE, C
C      RO: DENSITY OF WATER Kg/cu.m
C      RO = 1000.*(1.00076341 - 0.000076329*T -
&0.000003536*T**2)
RO = 1000.*(1.0031727496 - 0.00015900087*T -
&0.00000290393*T**2)
RETURN
END
```

10.2.12 SUBROUTINE BPE (TB, CB, BPR) to Compute the Boiling Point Elevation

```
c ++++ Boiling point evaluation(F)
SUBROUTINE BPE (TB, CB, BPR)
IMPLICIT DOUBLE PRECISION (A-Z)

C      TB: TEMPERATURE IN C
C      CB: CONCENTRATION MASS FRACTION
C      BPR: BOILING POINT RISE IN C
TK = TB + 273.
C = (19.819*CB)/(1. - CB)
DLOGTK = DLOG (TK)
BPR = (565.757/TK -9.81559 + 1.54739*DLOGTK
&      -(337.178/TK -6.41981 + 0.922753*DLOGTK)*C
&      + (32.681/TK - 0.55368 + 0.079022*DLOGTK)*C*C)
&      *(C/(266919.6/(TK*TK) - 379.669/TK +
&      0.334169))
RETURN
END
```

10.2.13 SUBROUTINE HCOEFF (TD, TF, CF, ID, L, W, A, FF, UO) to Compute the Heat Transfer Coefficient

```
SUBROUTINE HCOEFF (TD, TF, CF, ID, L, W, A, FF, UO)
C       TD: SATURATION TEMPERATURE C
C       TF: COOLING TEMPERATURE      C
C       L: TUBE LENGTH           M
C       ID: TUBE INSIDE DIAMETER    MM
C       W: FLOWRATE OF COOLANT      TON/min
C       A: HEAT TRANSFER AREA       SQ/M2
C       UO: HEAT TRANSFER COEFFICIENT    KCAL/min-M2 C
C       FF: FOULING FACTOR          min-M2-C/KCAL
IMPLICIT DOUBLE PRECISION (A-Z)
NTUBES = A*1000./(3.141592*ID*L)
FAREA = 3.141592*(ID/1000.)**2/4.*NTUBES
FLUX = W*1000./FAREA
TF1 = TF * 1.8 + 32.
TD1 = TD* 1.8 + 32.
CALL BRINRO (CF, TF, RO)
VM = FLUX/(RO*60.)
V = VM/0.3048
ID1 = ID/25.4
Z = 0.1024768E-02 - 0.7473939E-05*TD1
&      + 0.999077E-07*TD1**2 - 0.430046E-09*TD1**3
&      + 0.6206744E-12*TD1**4
Y = (V*ID1)**0.2/((160. + 1.92*TF1)*V)
UO1 = 1./(Z+Y)*0.081375
UO = UO1/(1. + FF *UO1)
RETURN
END
```

10.2.14 SUBROUTINE BRHI (C, T, NT, H, NH) to Compute Enthalpy of Brine (in Stages)

```
SUBROUTINE BRHI (C, T, NT, H, NH)
DOUBLE PRECISION C, T (NT), H (NH)
INTEGER NT, NH
DO     I = 1, NT
CALL BRH (C, T(I), H(I))
END DO
RETURN
END
```

10.2.15 SUBROUTINE TEXX (STAGE_NO, TB_IN, TB_OUT, T_EXX) to Compute the Nonequilibrium Temperature Difference for Brine

```
SUBROUTINE TEXX (STAGE_NO, TB_IN, TB_OUT, T_EXX)
C
C       Purpose: Compute non-equilibrium temperature difference
```

```
C       for brine
C
IMPLICIT NONE
C       Inputs
C
DOUBLE PRECISION STAGE_NO ! flash stage number
DOUBLE PRECISION TB_IN     ! brine inlet temperature,
C
DOUBLE PRECISION TB_OUT    ! brine outlet temperature,
C
C       Outputs
C
DOUBLE PRECISION T_EXX     ! non-equilibrium difference, degC
C
C       Internal variables
C
DOUBLE PRECISION T_DIFF
logical first time/.true./
If (first_time) then
 Type *, 'stage number is', stage_no
 first_time =.false.
endif
C
C Temperature rise
C
IF (TB_IN.GT. TB_OUT) THEN
 T_DIFF = TB_IN - TB_OUT
ELSE
T_DIFF = 0
END IF
C
C       Non-equilibrium temperature difference
C
IF (STAGE_NO.LT. 16.0) THEN
T_EXX = (39.032/TB_OUT - 5.679E-6*TB_OUT*TB_OUT
&                  + 0.0023*TB_OUT - 0.45)
&                                  *
(T_DIFF/4.5)**0.2
ELSE
T_EXX = 1.5229 - 0.0244 * TB_OUT
END IF

IF (T_EXX.LT. 0.0) THEN
T_EXX = 0.0
EISE IF (T_EXX. GT. 2.0) THEN
T_EXX = 2.0
END IF
RETURN
END
```

10.2.16 SUBROUTINE ORIFCD (WO, HO, WB, LEVEL, P_IN, P_OUT, RHO_B, CD, CC) to Calculate the Cd of the Brine Orifice

```
SUBROUTINE ORIFCD (WO,HO,WB,LEVEL, P_IN, P_OUT, RHO_B, CD, CC)
C
C       Purpose: Calculate Cd of brine orifice
C
IMPLICIT NONE
C
C       Inputs:
C
DOUBLE PRECISION WO          ! width of orifice, m
DOUBLE PRECISION HO          ! height of orifice, m
DOUBLE PRECISION WB          ! width of brine chamber, m
DOUBLE PRECISION LEVEL       ! brine level, m
DOUBLE PRECISION P_IN        ! upstream pressure, bar
DOUBLE PRECISION P_OUT       ! downstream pressure, bar
DOUBLE PRECISION RHO_B       ! brine density, kg/m3
C
C       Outputs
C
DOUBLE PRECISION CD          ! discharge coefficient
DOUBLE PRECISION CC          ! contraction coefficient
C
C       Internal variables
C
DOUBLE PRECISION R, CX
C
C       Parameters
C
DOUBLE PRECISION G, CH
PARAMETER (G = 9.81E-5)
PARAMETER (CH = 0.75)
C       Contraction parameter
IF (LEVEL.GT. HO) THEN
R = HO/(LEVEL + (P_IN-P_OUT)/(RHO_B*G))
ELSE
R = LEVEL/(LEVEL + (P_IN-P_OUT)/)RHO_B*G))
END IF

C       Coefficient of contraction (see "Fichtner - Handbook of
C       Seawater and Seawater Distillation")
CC = 0.61+ 0.18 * R - 0.58 * R*R + 0.7 * R*R*R
IF (CC.LT. 0.61) THEN
CC = 0.61
ELSE IF (CC.GT. 0.75) THEN
CC = 0.75
END IF

C       Coefficient of discharge
IF (LEVEL.GT. HO) THEN
```

```
CX = CC * CH * WO/WB * HO/LEVEL
ELSE
CX = CC * CH * WO/WB
END IF
CD = CC * CH/SQRT (1-CX*CX)
RETURN
END
```

10.2.17 SUBROUTINE HTCRIG (FB, TB, CB, FC, TC, FF, Di, L, t_w, N, U) to Calculate the Overall Heat Transfer Coefficient for the Condensing Tubes in the Flash Chambers and the Brine Heater

```
SUBROUTINE HTCRIG (FB, TB, CB, FC, TC, FF, Di, L, t_w, N, U)
C
C       Purpose:      To calculate the overall heat transfer
C                     coefficient for the condensing tubes in
the flash
C                     chambers and the brine heater.
C
C Comment: Computation corresponds to Plant design approach.
C       Put correct value of thermal conductivity of
C       tubes, add tube i.d./tube o.d. correction for
C       internal film coefficient.
C
IMPLICIT NONE
C
C       Arguments
C
DOUBLE PRECISION FB          ! Flow of brine in tubes (T/min)
DOUBLE PRECISION TB          ! Temperature in tubes (deg C)
DOUBLE PRECISION CB          ! Concentration of brine
                               (mass fraction)
DOUBLE PRECISION FC          ! Flow of condensate on tubes
                               (ton/min)
DOUBLE PRECISION TC          ! Temperature of condensate (C)
DOUBLE PRECISION FF          ! Fouling factor (m2-hr-C/kcal)
DOUBLE PRECISION Di          ! Tube internal diameter (mm)
DOUBLE PRECISION L           ! Length of tubes (m)
DOUBLE PRECISION t_w         ! Wall thickness (mm)
DOUBLE PRECISION N           ! Number of tubes
DOUBLE PRECISION U           ! Overall htc (kcal/min-m2-C)
C
C       Internal variables
C
DOUBLE PRECISION visc_w, Rel, visc_b, Re
DOUBLE PRECISION cp_b, k_b, Pr
DOUBLE PRECISION hi, ho
DOUBLE PRECISION wc, visc_c, k_c
DOUBLE PRECISION FF_SI
```

```
DOUBLE PRECISION t_w_m, Wr
DOUBLE PRECISION U_SI
C
C       Parameters
C
DOUBLE PRECISION PI, g, rho_c, k_t
DOUBLE PRECISION J_per_kcal, sec_per_hr, sec_per_min
DOUBLE PRECISION mm_per_m, kg_per_ton, SI_to_cP
PARAMETER (J_per_kcal = 4184)
PARAMETER (sec_per_hr = 3600)
PARAMETER (sec_per_min = 60)
PARAMETER (mm_per_m = 1000)
PARAMETER (kg_per_ton = 1000)
PARAMETER (SI_to_cP = 1000)
PARAMETER (pi = 3.141592654)
PARAMETER (g = 9.81)                    ! (m/s2)
PARAMETER (rho_c = 1000)                ! (kg/m3)
PARAMETER (k_t = 290                    ! thermal cond'ty of tubes
&       * J_per_Kcal/sec_per_hr)        ! (kcal/m-h-C-  > W/m-K)
C
C       Note:All intermediate calculations are performed in SI
C       units (kg, m, s).
C       The OHTC is converted at the end into units of
C       kcal/min-m2-C.
C
C       1) TUBE INSIDE HTC
C
C       Reynolds number
visc_w = (7.15E-5 * TB*TB - 0.01611 * TB + 1.1854)/
&       SI_to_cP
Rel = 0.968+ 3.3E-4 * TB + 2.8 * CB + 1.092E-3 * TB * CB
visc_b = Rel * visc_w
Re = (4 * FB/N * kg_per_ton/sec_per_min)
&       /(pi * Di/mm_per_m * visc_b)
C       Prantl number
cp_b = (0.988 + 1.5E-4*TB - 1.0*CB + 1E-3*CB*TB) *
&       J_per_kcal
k_b = 0.52 + 2.4E-3 * TB - 2.4E-5 * TB * TB
Pr = cp_b * visc_b/k_b
C       Tube inside htc
hi = (0.022 * Re**0.82 * Pr**0.4) * k_b/(Di/mm_per_m)
C
C       2) TUBE OUTSIDE HTC
C
C       Condensate rate per unit length
wc = (FC * kg_per_ton/sec_per_min)/(L * N)
C       Viscosity of water
visc_c = (7.15E-5 * TC*TC - 0.01611 * TC + 1.1854)/
&       SI_to_cP
C       Thermal conductivity of condensate
```

```
k_c = 0.577 + 1.522E-3 * TC - 5.81E-6 * TC*TC
C        Tube outside htc
ho = 1.89 * k_c * (g * rho_c*rho_c/(4 * wc * visc_c))
&        ** 0.3333
ho = 0.39685 * ho
C
C        3) FOULING FACTOR
C
C        Convert units to SI, from m2-hr-K/kcal
FF_SI = FF * sec_per_hr/J_per_kcal
C
C        4) THERMAL RESISTANCE OF TUBE WALL
C
C        Wall thickness in meters
t_w_m = t_w/mm_per_m
C        Wall resistance
Wr = t_w_m/k_t
C
C        THUS WE GET THE OHTC IN W/M2 - K:
C
C        First in SI (add correction for tube i.d./tube o.d.)
U_SI = 1.0/(1/hi* (Di+2*t_w)/Di+ Wr + FF_SI + 1/ho)
C        Then in kcal/min-m2-C
U = U_SI/J_per_kcal * sec_per_min
C
C        ALL DONE
C
RETURN
END
```

10.2.18 SUBROUTINE VAPH (TS, PS, H) Calculates the Enthalpy of Water Vapor as a Function of the Temperature and Pressure

```
SUBROUTINE VAPH (TS, PS, H)
C- - - - - - - - - - - - - - - - - - - - - - - - - - - - - - - - -
C      This routine calculates the enthalpy of water Vapor
C      as a function of Temperature and Pressure
C
C      Required input : ts - Temperature of Steam, C
C                       ps - Pressure, BAR
C            Output : h - Enthalpy, Kcal/kg
C
C      <<<< Reference: PERRY'S Chemical Engineering
C      HandBook >>>>>>
C- - - - - - - - - - - - - - - - - - - - - - - - - - - - - - - - -
P = PS*0.9869
T = TS+273.0
Tau = 1./T
Tau2 = Tau*Tau
Tau3 = Tau2*Tau
```

```
B0 = 1.89 - 2641.62*Tau*(10.0)**(80870.0*Tau2)
BD0 = - 2641.62*(10)**(80870.0*Tau2)*(372420.11*Tau2+1.0)
G1 = Tau*(82.546 - 1.6246E5*Tau)
DG1 = 82.546 - 3.2492E5*Tau
G2 =.21828 - 1.2697E5*Tau2
DG2 = -2.5394E5*Tau
G3 = 3.635E-4-6.768*(Tau3*1.0E8)**8
DG3 = -162.432*(Tau3*1.0E8)**8/Tau
F0 = B0+Tau*DB0
F1 = B0*B0*Tau* (2.0*G1+Tau*DG1) +2.0*Tau2*B0*G1*DB0
F3 = (B0*Tau) * (B0*Tau) * (B0*Tau) * (4.0*B0*G2+Tau*B0*DG2 +
&       4.*Tau*G2*DB0)
F12 = (Tau*B0) **12* (13.*B0*G3+Tau*B0*DG3+13.*Tau*G3*DB0)
F12 = -F12
FP = (1.0E4/1012.95)*(1.472*(T-273.16) +
&       3.7783E-4*(T*T-74516.386) + 47.8365*ALOG(T/273.16) +
2502.36)
HC = P*(F0+P*(F1/2.0+P*P*(F3/4.+F12*P**9/13.))) +FP
C
H = 0.0242*HC
RETURN
END
```

10.3 MATLAB Programs

The following source programs are used with Standard MATLAB PROGRAM routines in the computation of the optimal PID controller parameters for the well-known *first order plus dead time* (FODT) normalized form of plant model. These results are fitted into simple formulae, which can be used as readymade optimal tuning rules. The optimal design is carried out with a wide variety of integral performance criteria such as ISE, ISTE, and IAE, ITAE.MATLAB routines are used PID controller design methods. Removal of the need to approximate the plant or model with dead-time (FODT) form is a significant step.

The plant model information is needed. The proposed method of optimal controller design is of significance to practical applications.

10.3.1 Controller Tuning and Process Identification: Criteria for Good Control: ISE, IAE, ITAE

```
% Controller Tuning and Process Identification
% Criteria for good control: ISE IAE ITAE
%
% [EH XOUT] = PROCES(X0, ALGORITHM)
%
% ALGORITHM : 1- FMINS     (default)
```

```
%                  2- FMINU
%        TYPE : 1- ISE 2- IAE 3- ITAE
%
% REQUIRED INPUT:
% T1   : the time sequence associated with response
% X0   : [Kc Ti Td]
% Kc   : Proportional gain
% Ti   : Integral time
% Td   : Derivative time
% PP   : Impulse response of the process
% TT   : Sample time
%
% Global declaration PP TT T1 TYP
%
% OUTPUT:
%      XOUT : Optimum parameter for {kc ti td}
%      EH : Performance criterion ISE IAE ITAE
% see also TRAPID
%
Function [eh,xout] = proces(x,mode)
option = zeros(1,14);option(1) = 1;option(2) = 1e-03;option(3)
= 1e-04;

if nargin = = 1, mode = 1; end
if mode = = 1
xout = fmins('trapid',x,option);
else
xout = fminu('trapid',x);
end
[eh] = trapid(xout);
End
```

10.3.2 PID Control Simulation Using a Time-Domain Description of the Subsystems with a Recursive Trapezoidal Integration

```
% PID control simulation using time-domain description of the
% subsystems
%
% Input: Kc = Proportional gain
%        Ti = Integral Time
%        Td = Derivative time
%        T = Sampling time
%        r = setpoint
%        p = impulse response of the process
%        x = [kc, Ti, Td]
%
% Output: y = measured value
%         e = error
%         u = controller output
```

```
%
Function [y,u,e] = Pid_sim(x,p,t)
%
[m,n] = size(p);
se = 0;
nc = max([m,n]);
if m = =1, p = p'; end
r = [0;ones(nc-1,1)];
kc = x(1); ti = x(2); td = x(3);
a = t/2*p(1)*kc(1+0.5*t/ti + td/t);
%
y(1)  = (a*r(1))/(1 + a);
e(1)  = r(1) - y(1);
u(1)  = kc*(1 + 0.5*t/ti + td/t)*e(1);
%
y(2)  = (0.5*t*p(2)*u(1)+0.5*t*p(1)*kc*(0.5*t/ti-
td/t)*e(1)+a*r(2))/(1+a);
e(2)  = r(2) - y(2);
u(2)  = kc*(e(2) + (0.5*t/ti)*(e(1)+e(2))+td/t*(e(2)-e(1)));
%
For k = 3:nc
se = se +e(k-1);
spu = 0;
for j = 2:k-1
spu = spu + (p(k-j+1)*u(j));
end
spu = 0.5*t*p(k)*u(1) + t*spu;
y1 = 0.5*t*kc*p(1)*(0.5*t/ti*e(1) - td/t*e(k-1) +t/ti*se);
y2 = 0.5*t*kc*p(1)*(1+0.5*t/ti+td/t)*r(k);
%
y(k)  = (spu + y1 + y2)/(1+a);
e(k)  = r(k) - y(k);
u(k)  = kc*(e(k) + 0.5*t/ti*(e(1)+e(k))+t/ti*se+td/t*(e(k)-
e(k-1)));
end
end
```

10.3.3 PID Control Simulation Using a Time-Domain Description of the Subsystems with Recursive Trapezoidal Integration Including the Calculation of Performance Indices

```
% PID control simulation using time-domain description of the
% subsystems including calculation of performance indices with
% recursive trapezoidal integration
%
% Input: The controller parameters :
% kc = Proportional gain; Ti = Integral time
% Td = Derivative time; T = Sampling time
% r = setpoint; p = impulse response of the process
% x = [kc, Ti, Td] T0 = time vector
```

```
% Output: y = measured value; e = error; u = Controller output
%
% Global declaration pp t1 tt typ
%
%        See also: PROCES
%
Function [eh] = trapid(x,mode);
p = pp; t = tt; t0 = t1;[m,n] = size(p);nc = max([m,n]);
if m = = 1, p = p'; end
kc = x(1); ti = x(2); td = x(3);
r = [zeros(1,1); ones(nc-1,1)]; se = 0;
a = t/2*p(1)*kc*(1 + 0.5*t/ti + td/t);
%
y(1) = (a*r(1))/(1 + a);
e(1) = r(1) - y(1);
u(1) = kc*(1 + 0.5*t/ti + td/t)*e(1);
%
y(2) = (0.5*t*p(2)*u(1)+0.5*t*p(1)*kc*(0.5*t/ti-td/t)*
e(1)+a*r(2))/(1+a);
e(2) = r(2) - y(2);
u(2) = kc*(e(2) + (0.5*t/ti)*(e(1)+e(2))+td/t*(e(2)-e(1)));
%
for k = 3:nc
se = se + e(k-1); spu = 0;
for j = 2:k-1
spu = spu + (p(k-j+1)*u(j));
end
spu = 0.5*t*p(k)*u(1) + t*spu;
y1 = 0.5*t*kc*p(1)*(0.5*t/ti*e(1) - td/t*e(k-1) + t/ti*se);
y2 = 0.5*t*kc*p(1)*(1 + 0.5*t/ti + td/t)*r(k);
%
y(k) = (spu + y1 + y2)/(1+a);
e(k) = r(k) - y(k);
u(k) = kc*(e(k)+0.5*t/ti*(e(1)+e(k))+t/ti*se+td/t*(e(k)-
e(k-1)));
end
if typ = = 1,
ise = t*(e*e');
eh = ise;
elseif typ = = 2,
iae = t*(sum(abs(e)));
eh = iae;
elseif typ = = 3,
itae = abs(e)*t0;
eh = itae;
elseif typ = = 4,
ee = (t0.*e'); iste = (e*ee);
eh = iste;
end
end
```

10.3.4 PROCESID: Time-Domain Process Identification

```
% PROCESID (TIME DOMAIN PROCESS IDENTIFICATION PROGRAM)
% THIS PROGRAM REQUIRES STEP (RESPONSE) TEST DATA INPUT AND
WILL
% RETURN THE PARAMETERS OF A FIRST ORDER WITH DEADTIME OR A
% SECOND ORDER WITH DEADTIME PROCESS MODEL
% Note: This program uses a function called FMINS that
% implements the Nelder-Mead Simplex algorithm for minimizing a
% nonlinear function of several variables; or FMINU that uses
% quasi-Newton method.
%
% [YHAT,X,MSE] = PROCESID(X0,ALGORITHM)
% The transfer function of the system is:
% TF = K.e(-THETA.s)/(1+TAU.S)      (1st order)
% OR
% TF = K.e(-THETA.s)(1+TAU.S)/[(1+TAU1.S)(1+TAU2.S)] (2nd
order)
% YHAT: the step response of the fitted 1st/2nd order model
% X(1st order):estimated (THETA,TAU)            (known gain)
% X(1st order):estimated (K,THETA,TAU)   (unknown gain)
% X(2nd order):estimated (K,THETA,TAU,TAU1,TAU2)   (known gain)
% X(2nd order):estimated (K,THETA,TAU,TAU1,TAU2) (unknown
gain)
% MSE: the mean-square-error
%
% X0 : initial values for X defined as above
% ALGORITHM: 1. FMINS
%       2. FMINU
% Required input data:
%       YY: the step response of the identified plant
%       TT: the time sequence associated with YY
%       KK: the steady state gain of the plant (if known)
%
%       see also PROCFUN
%
FUNCTION [YHAT, XOUT, MSE] = POCESID(X,MODE)
IF NARGIN = = 1, MODE = 1; END
K = KK; Y = YY; T = TT;
OPTION = ZEROS(1,14); OPTION (1) = 0;OPTION (2) =
1E-04;OPTION(3) = 1E-04;
%
IF LENGTH(X) < 4
%FIRST ORDER WITH DEAD TIME MODEL
IF LENGTH(X) = = 2
%FINDS THETA, TAU
IF MODE = = 1
XOUT = FMINS('PROCFUN',X,OPTION);
ELSE
```

```
XOUT = FMINU('PROCFUN',X);
END
THETA = XOUT(1);TAU = XOUT(2);
ELSE
% FIND K, THETA, TAU
IF MODE = = 1
XOUT = FMINS('PROCFUN',X,OPTION);
ELSE
XOUT = FMINU('PROCFUN',X);
END
K = XOUT(1);THETA = XOUT(2);TAU = XOUT(3);
END
ELSE
%SECOND ORDER WITH DELAY
IF LENGTH(X) = = 4
%FIND THETA, TAU, TAU1, AND TAU2
IF MODE = = 1
XOUT = FMINS('PROCFUN',X,OPTION);
ELSE
XOUT = FMINU('PROCFUN',X);
END
THETA = XOUT(1); TAU = XOUT(2); TAU1 = XOUT(3); TAU2 =
XOUT(4);
ELSE
%FIND K, THETA, TAU, TAU1 AND TAU2
IF MODE = = 1
XOUT = FMINS('PROCFUN',X,OPTION);
ELSE
XOUT = FMINU('PROCFUN',X);
END
K = XOUT(1);THETA = XOUT(2);TAU = XOUT(3);TAU1 = XOUT(4);
TAU2 = XOUT(5);
END
END
MSE,YHAT] = PROCFUN(XOUT);
PLOT(T,[Y,YHAT]),GRID
END
```

10.3.5 PROCFUN: Computation of the Step Response of the First-/Second-Order Transfer Function with Delay

```
% PROCFUN computes the step response of 1st/2nd order transfer
% function with delay. The transfer function is given by the
% equations:
% (1) TF = e^(-THETA.s).K/(1+TAU.S)        1ST ORDER
% (2) TF = e^(-THETA.s).K(1+TAU.S)/[(1+TAU1.S) (1+TAU1.S)] 2nd
order
% [MSE,YSTEP] = PROCFUN(X)
```

```
% YSTEP : THE COMPUTED STEP RESPONSE
% X : [K, THETA, TAU]        {FIRST ORDER}
%       [K, THETA, TAU, TAU2] {SECOND ORDER}
% T : TIME VECTOR
%
FUNCTION(MSE,YSTEP) = PROCFUN(X);
T = TT; Y = YY; K = KK; N = LENGTH(T);
DT = T(2)-T(1); SLOPE = -1/(4*DT);PENALTY = 1;
SIZE(T); IF ANS(1) = = 1,TW = T'; END
IF LENGTH(x) = = 2,X = [K X]; END
%
K = X(1);THETA = X(2); TAU = X(3);
ND = FIX(THETA/DT + 0.999999); W = ONES(TW);
IF ND > 0; W = ONES(N-ND,1);TW = TW(ND+1:N);END
IF LENGTH(x) < 4
% FIRST ORDER WITH DELAY
Y1 = K*W - K*EXP(-TW/TAU+THETA/TAU*W);
ELSE
% SECOND ORDER WITH DELAY
TAU1 = X(4);TAU2 = X(5);
IF TAU1 ~ = TAU2
a = (TAU-TAU1)/(TAU1-TAU2); b = (TAU-TAU2)/(TAU2-TAU1);
Y1 = K*(W + a*(EXP(-TW/TAU1+W*THETA/TAU1))+b*(EXP(-TW/TAU2 +
W*THETA/TAU2)));
ELSEIF TAU1 = = 0
IF TAU = = 0
Y1 = K*W;
ELSE
'CASE IS PHYSICALLY UNREALIZABLE'
END
ELSE
V = (TW - THETA*W)/TAU1;
Y1 = K*(W-EXP(-V).*(W+(1-TAU/TAU1)*V));
END
END
IF ND > 0
YSTEP = [ZEROS(ND,1);Y1];
ELSE
YSTEP = Y1;
END
IF THETA < 0, PENALTY = 1.2 + THETA*SLOPE;
END
ER = Y - YSTEP; MSE = (ER'*ER/(LENGTH(Y1)))*PENALTY;
END
```

Glossary

Action mode: The combination of the three (proportional, integral, and derivative) elements that are kept in action when the PID controller is applied.

Adaptive control: This is a control system in which the controller is automatically adjusted to compensate for changing process conditions.

Analytical model: This is based on the physical principles involved in a certain process.

Antiwindup: A mechanism to prevent windup of the integral element.

Artificial neural network: A set of interconnected artificial neurons that can perform a simple or complex task.

Artificial neural networks: These consist of numerous simple processing units called *neurons* that are linked via weighted connections.

Artificial neuron: The smallest information processing unit in an artificial neural network, which emulates a simple model of a biological neuron.

Backpropagation: A learning procedure for multilayer networks in which error signals are propagated in reverse direction.

Biofouling: The formation of bacterial film (biofilm) on fragile reverse osmosis membrane surfaces.

Black-box model: The unknown parameters in a selected model are determined through experiments. The technique is known as identification.

Black-box modeling: Type of modeling in which the system is considered absolutely unknown and no information based on the laws of physics is used to construct the model.

Blowdown: Waste from a wet cooling tower—this water will have been cycled as many times as possible and will have reached the maximum allowable (and safe) limits of certain dissolved solids.

Blowthrough: In the MSF plant, the vapor generated in each flash stage is supposed to flow upward toward the condenser. However, when the brine level in the stage becomes low enough the vapor leaks to the successive stage. This is known as vapor blowthrough.

Boiling point: The temperature at which a liquid's vapor pressure equals the pressure acting on the liquid.

Boiling point elevation (BPE): The difference between the boiling point of a solution and the boiling point of pure water at the same pressure.

Boiling point rise (BPR): See "boiling point elevation."

Brackish water: Water containing a low concentration of soluble salts, usually between 1000 and 10,000 mg/L.

Brine: Water saturated with, or containing a high concentration of salts, usually in excess of 36,000 mg/L.

Brine concentrator: Term used to describe a vertical tube falling film evaporator employing special scale control techniques to maximize the concentration of dissolved solids.

Brine heater: The heat input section of a multi-stage flash desalination plant where feedwater is heated to the process' top temperature.

CACSD: Computer-aided control systems design.

Closed-loop control system: Also called a *feedback control system*. This is a control system, in which the difference between outputs and setpoints is used by the control law, in order to calculate the control action.

Concentrate: Water that contains a high concentration of salt. Concentrate discharges from desalination plants may include constituents used in pretreatment processes, in addition to the high salt concentration seawater.

Concentration: (1) The amount of a substance dissolved or suspended in a unit volume of solution. (2) The process of increasing the amount of a substance per unit volume of solution.

Condensate: Water obtained by evaporation and subsequent condensation.

Condensation: The change in state from vapor to liquid; the opposite of evaporation.

Condenser: A heat exchanger device used to cool steam and convert it from the vapor to the liquid phase.

Constraints: Restrictions placed over the process parameters such as on pressure, temperature, flowrate, and so on, not to exceed their lower and upper bounds are inequality constraints. Moreover, there are equality constraints that describe mass, energy, and momentum balances.

Control law: This is the procedure implemented in the controller to generate the manipulated variables.

Control system: A control system is an interconnection of components that act together to satisfy a common control objective.

Controlled variable (output): This is a variable in the control system that the system tries to keep under control, that is, by trying to keep it constant or by having it follow an assigned reference variable.

Controller: This is a control device that receives information from the measuring devices and, after calculations, decides what action should be taken. It is the decision maker that implements the control law.

Critical flow: The flow of any liquid, such as brine, is termed critical when the dimensionless Froude (Fr) number ($Fr = u/\sqrt{gl}$) is equal to zero, where u is the flow velocity (m/s), g is the acceleration due to gravity, and L is the brine level at the vena contracta (m).

DAE: A system of equations in which some are differential equations and the remaining are algebraic equations.

Deaerator: A process unit that removes the dissolved air and other gases from the brine.

Degree of freedom: This is the difference between the total number of process variables and the number of independent equations defined in the model.

Demister: A strainer, or fine-steal mesh, inside the flash chamber, designed specifically to stop the carryover of liquid droplets, containing salts, from going through into the vapor area.

Desalination: The process of removing salt from water as total dissolved solids (TDS), typically performed by either reverse osmosis (RO), multi-stage flash (MSF), or multieffect desalination (MED).

Desuperheater: This is a process unit in which superheated steam is mixed with steam condensate to reduce or remove its superheat and, thus, change it into saturated steam. Saturated steam has only one degree of freedom, either pressure or temperature, whereas in the case of superheated steam both temperature and pressure are independent.

Distillation: A process of desalination where the intake water is heated to produce steam. The steam is then condensed to produce product water with low salt concentration.

Disturbance: Any deviation in the fixed input conditions such as temperature, pressure, or composition entering the process is known as a disturbance.

Drinking water: Water safe for human consumption or which may be used in the preparation of food or beverages, or for cleaning articles used in the preparation of food or beverages.

Dynamic network: A network capable of learning dynamic or temporal behaviors.

Dynamical model: This model describes the temporal evolution of the process. In a mathematical representation, dynamical models involve differential equations.

Economic water scarcity: Economic water scarcity is a term describing a region that has adequate physical water resources to meet their water supply needs, but must increase the availability of the water through additional storage and conveyance facilities. Most of these countries face severe financial and development capacity problems for increasing the primary water supply by building the needed infrastructures.

Effluent water: Water that flows from a sewage treatment plant after it has been treated.

Electrodialysis: An electrochemical separation process in which ions are transferred through anion- and cation-selective membranes from a less concentrated to a more concentrated solution as a result of the passage of a direct electric current. This kind of desalination technology is used mostly for brackish water.

Evaporation: The process in which water is converted to a vapor that can be condensed.

Exactly specified process: This is a process whose model has no degrees of freedom.

Expert systems: This is a program system, which represents and applies factual knowledge of specific areas of expertise to solve problems.

Feedforward neural network: A multilayer network in which information flows only in a forward direction.

Feedwater: Water fed to desalination equipment. This can be source water with or without pretreatment.

Filtration: A process that separates small particles from water by using a porous barrier to trap the particles and allowing the water through.

Flash distillation: See "multi-stage flash desalination."

Flash evaporator: A distillation device where saline water is vaporized in a vessel under vacuum through pressure reduction. See also "multi-stage flash desalination."

Flashdown: The difference between the flashing temperature of the first and the last stages in the MSF plant is called a *flashdown*.

Flashing: The process of vaporizing a fluid by pressure reduction rather than temperature elevation.

Flowsheet: A diagram that shows interconnections between various process units in a production plant and indicates directions of mass and energy flows through stream.

Freshwater: Water that contains less than 1000 milligrams per liter (mg/L) of dissolved solids; generally, more than 500 mg/L of dissolved solids is undesirable for drinking and many industrial uses.

Hydraulic jump: The flashing brine flowing in a stage takes a sudden jump due to a restriction placed in its flow path, thus increasing the brine level near the orifice wall. It is known as a hydraulic jump.

IAE: Integral absolute error.

IMC: Internal model control.

Index: Minimum number of differentiations that are to be carried out to convert a DAE system into a system of only differential equations.

Interpolation—linear: Estimating intermediate values between two specified or measured values of a parameter is known as interpolation. If a straight-line relationship is assumed between the given values, it is called a *linear interpolation*.

IS: Internal stability.

ISE: Internal square error.

ITAE: Integral time-weighted absolute error.

ITSE: Integral time-weighted square error.

Kick-plate: Type of restriction placed in the flow path of the flashing brine due to which a hydraulic jump occurs.

Learning algorithm: A set of rules, which are applied during the training phase to adjust the parameters of a neural network in order to make it perform better.

Local model network: A network that decomposes the input space in different local regions to find a local model for each region.

Logarithmic mean temperature difference: In a heat exchanger, when the difference between the temperature of hot and cold streams at one end is ΔT_1 and at the other end ΔT_2 the logarithmic mean temperature difference is $\Delta T_{lm} = (\Delta T_1 - \Delta T_2)/\ln (\Delta T_1/\Delta T_2)$.

Mass transfer: This is a phenomenon in which a certain mass is transferred between two different phases, for example, from a liquid surface into a vapor phase. Such transfer occurs due to the difference in fugacities or simply vapor pressures of the transferring component between the two phases.

Matrix: If various quantities pertaining to a process are arranged in rows and columns, it gives rise to a process matrix.

Matrix–sparse: If a process matrix contains many quantities equal to zero, it is called a *sparse matrix*.

Matrix–tridiagonal: A sparse matrix in which elements are only available on three diagonals.

MGD: Abbreviation for million gallons per day. This term is used to describe the volumes of water treated and discharged from a treatment plant.

Microfiltration: A physical separation process where tiny, hollow straw-like membranes separate particles from water. It is used as a pretreatment for reverse osmosis.

MIMO: Multi-input, multi-output.

Model: Describes the behavior of a system from a particular point of view.

Model—dynamic: A dynamic model represents the transient condition of a process plant, with time as the independent variable.

Model—first principle based: A model that is based on the physical principles involved in the process.

Model—linear/nonlinear: A model containing only linear equations is termed a *linear model*; the equations are represented by straight lines. On the other hand, model parameters or variables having powers exceeding one give rise to *nonlinear models*.

Model—parametric: An empirical model that does not involve any physical considerations. It is developed by collecting actual data and suitably fitting or correlating the data.

Model—steady state: Such a model represents steady-state operation of a process plant, in which time is not among the independent variables.

Model predictive control: Plant control is implemented in combination with a process model, which predicts the output for the given input to the plant.

Model process: Adequate mathematical description of the working of a process plant is its model.

MSF: See "multi-stage flash desalination."

Multieffect desalination (MED): The operation of an evaporator in which brine or seawater is evaporated on a large scale under vacuum by passing the water over hot metal surfaces heated by steam in internal tubes. Multiple effect distillation is similar to MSF because it takes place in a series of containers (called *effects*), each at a lower pressure than the previous one, and utilizes the processes of evaporation and condensation. An effect is made up of a container and a heat exchanger. Some of the feedwater in each effect boils, producing steam. The steam condenses, giving rise to freshwater, and the condensation releases heat to evaporate water in the next effect. Several evaporators are used in series, and the whole system is called a *multiple effect boiling* (MEB or MED) plant.

Multilayer perceptron: A network consisting of neurons with nonlinear activation functions arranged in more than one layers.

Multi-stage flash desalination (MSF): A process unit in which brine or seawater is flashed into vapor on a large scale. In multi-stage flash desalination, seawater is heated in a container called the *brine heater*. The heated water then flows to a second container, called the *stage*, where the pressure is lower. The lower pressure causes a portion of the water to boil because at a lower pressure, water has a lower boiling temperature. In fact, the water boils so quickly it is said to "flash" into steam. The remaining water is then moved to the next stage, where the pressure is even lower, causing more water to flash into steam. It consists of a flash chamber, demister, overhead condenser, and a tray to receive salt-free water formed by condensation of vapor. When several stages are interconnected, it makes a multistage flash (MSF) plant that is either operated as once-through or with brine recirculation.

Newton–Raphson method: A numerical method that is applied to solve an equation of the type $f(x) = 0$. Starting with a known approximation x_i, another approximation x_{i+1} is calculated from $x_{i+1} = x_i - f(x_i)/f'(x_i)$, where f' is the derivative of f. The procedure is continued until the aforementioned equation is satisfied.

NN: Neural network.

Noncondensables: Gases such as air and carbon dioxide that do not condense to a liquid with the water vapor when heat is removed.

Nonequilibration: The difference in brine temperature and the temperature that can be attained in thermal equilibrium.

Nonequilibrium allowance: The flashing brine flowing through a particular stage having finite residence time does not come to a thermal equilibrium state at its exit. Therefore, a nonequilibrium allowance is used in calculations to account for its deviation from the equilibrium state.

Online optimization: Process optimization is performed in a working plant by collecting operation data and feeding the same to a model, which calculates optimal parameters for control.

Optimization: It aims to find the best operating conditions for a production plant, for example, those at which the water production rate is maximum or those for achieving maximum performance ratio (*pr*).

Parameter estimation: Estimation of the unknown parameters of the selected model.

Parts per million: A unit used to measure contamination concentration (parts of contamination per million parts of water). One part per million is equal to 1 mg/L. (This term is becoming obsolete as instruments measure smaller particles.)

Performance ratio: A unit of measurement used to characterize evaporator performance, expressed as the mass of distillate produced per unit of energy consumed.

Performance simulation: Simulation that is carried out for the given inputs to compute performance criteria of a plant, such as its production rate or product purity.

Physical water scarcity: Physical water scarcity is a term used to describe an area whose primary water supply is developed at 60% or greater than the total potential capacity. One must understand that the total potential capacity includes water that can never be entirely accessed. These countries do not have sufficient freshwater to meet their demands for agriculture, domestic water, industrial sectors, and environmental requirements.

PID: Proportional, integral, and derivative.

PID control: The feedback control method that uses the PID controller as its main tool.

PID controller: A controller consisting of the proportional, integral, and derivative elements.

Posttreatment: The treatment processes following desalination, usually employed to stabilize water and reduce its corrosivity and improve its taste.

Potable water: Water that does not contain pollutants, contamination, objectionable minerals, or infective agents and is considered safe for domestic consumption; drinkable. See also "drinking water."

PPM: See "parts per million."

Process: A system that produces certain product(s) in their widest sense.

Process model: A mathematical description that expresses the characteristics of a process.

Recurrent network: A network with internal feedback paths.

Reverse osmosis (RO): A method of removing salts or other impurities from water by forcing water through a semipermeable membrane.

Saline: Containing or resembling sodium chloride or similar salts.

Saline water: Water that contains significant amounts of dissolved solids. Parameters for saline *water*: *Freshwater*—less than 1000 parts per million (ppm). *Slightly saline water*—from 1000 to 3000 ppm. *Moderately saline water*—from 3,000 to 10,000 ppm. *Highly saline water*—from 10,000 to 35,000 ppm.

Salinity: Generally, the concentration of mineral salts dissolved in water. Salinity may be measured by weight (total dissolved solids [TDS]), electrical conductivity, or osmotic pressure. Where seawater is known to be the major source of salt, salinity is often used to refer to the concentration of chlorides in the water.

Salt: A class of ionic compounds formed by the combination of an acid and a base, of which sodium chloride is one of the most common examples.

Scaling: Salt deposits on the interior surfaces of a desalination plant.

Seawater: General term for sea or ocean water, with a typical total dissolved solids concentration of 35,000 mg/L.

Self-organizing learning: A type of learning in which the neural network organizes itself without a supervisor.

Setpoints: Values that are specified and set in the controllers for controlling different parameters, for example, the top brine temperature (TBT) in the MSF plant.

Simulation: Using a process model, calculating outputs for given inputs, is known as simulation.

Simulator: This is a package of software, possibly with hardware components, capable of simulating plant performance under different sets of conditions. It is used for training plant operators and can be helpful in designing extensions or new plants.

SISO: Single-input, single-output.

Solar still: A simple device for evaporating and condensing water using only solar energy in order to provide a supply of potable water.

SPEEDUP: Simulation program for evaluation and evolutionary design of unsteady processes from Aspen Technology, Cambridge, MA, USA. It is an equation-based flowsheeting package.

Splitter: A unit that divides a process stream into at least two streams. The split streams as well as the parent stream have the same composition and other properties.

Stage: One of several units of a flash evaporator, each of which operates at a successively lower pressure.

Standard seawater: A widely accepted "standard" total dissolved solids concentration of approximately 36,000 mg/L, considered to be typical of most seawaters.

Startup/shutdown: These are the procedures that are followed for starting up a plant from a nonworking condition and bringing it to a steady-state operation or conversely shutting down operations of a plant from full stream conditions. Both are transient in nature and would be represented by dynamic models.

Stream: The stream variables relate to the process unit flow variables.

Structure optimization: Optimization of the structure of the model. For example, in the case of a neural model, structure optimization includes determination of proper signal delays, number of neurons in each layer, or number of hidden layers.

Subcritical/supercritical flow: Subcritical flow occurs when the Froude number $Fr < 1$ and supercritical when $Fr > 1$.

Submerged flow: The flow is termed as submerged when the brine level at the vena contracta and the level at the end of the flash stage are not very different.

Sustainable development: Development that meets the needs of the present without compromising the ability of future generations to meet their own needs.

Systems identification: The process of determination of a model of an unknown dynamic system.

TBT: See "top brine temperature."

TDS: Total dissolved solids. A quantitative measure of the residual minerals dissolved in water that remain after evaporation of a solution, usually expressed in milligrams per liter.

Thermodynamic: A means of converting heat into mechanical work.

Top brine temperature (TBT): The maximum temperature of the fluid being evaporated in an evaporator system.

Tuning: The engineering work to adjust the parameters of a PID controller so that the control system exhibits a desired property.

Two-degrees-of-freedom PID controller: The modern type of PID controller, which can adjust two closed-loop transfer functions separately.

Vapor: The gaseous phase of a material that is in the solid or liquid state at standard temperature and pressure.

Vapor compression: A desalination process in which seawater is evaporated and the vapor is compressed. Mechanical or thermal energy is used to compress the vapor, which increases its temperature. The vapor is then condensed to form product water and the released heat is used to evaporate the seawater.

Vena contracta: When a liquid stream passes through a constriction, such as an orifice, its cross section becomes narrower than the orifice diameter. The narrowest cross section past the orifice is called the *vena contracta*.

Water purification: The process of removing undesirable chemicals, biological contaminants, and materials from water so that is becomes safe to use. Water quality—a term used to describe the chemical, physical, and biological characteristics of water, usually in respect to its suitability for a particular purpose.

Bibliography

Al-Gobaisi, D.M.K.F. (1991), An overview of modern control strategies for optimizing thermal desalination plants, *Desalination*, **84**, 3–43.

Al-Gobaisi, D.M.K.F. (1994), A quarter-century of seawater desalination by large multistage flash plants in Abu Dhabi, *Desalination*, **99**, 483.

Al-Gobaisi, D.M.K.F. (2001), Overview of desalination and water resources, in *Encyclopedia of Desalination and Water Resources (DESWARE)*, EOLSS Publishers, Oxford, UK (www.desware.net).

Al-Gobaisi, D.M.K.F., A. Hassan, A.S. Barakzai, and M.A. Aziz (1992), Manageable automation system for power and desalination plant, in *Proceedings of DESAL'92*, Al-Ain, United Arab Emirates, pp. 773–814.

Al-Gobaisi, D.M.K.F., A. Hassan, A. Woldai, G.P. Rao, and R. Borsani (1993), Towards improved automation for desalination process, part I: Advance control, in *IDA and WRPC World Conference on Desalination and Water Treatment*, Yokohama, Japan, November 3–6, 1993, Vol. III, pp. 57–94.

Al-Gobaisi, D.M.K.F., A. Husain, G.P. Rao, A. Woldai, and R. Borsani (1994), Toward improved automation for desalination processes, part I: Advanced control, *Desalination*, **97**, 469.

Al Gobaisi, D.M.K.F., B. Makkawi, and A.M. El-Nashar (2010), Renewable energy versus nuclear energy, in *International Conference on Renewable and Alternative Sources of Energy*, Beirut, Lebanon, November 25–26, 2010.

Åstrom, K.J., T. Hagglund, C.C. Hang, and W.K. Ho (1993), Automatic tuning and adaptation for PID controllers—A survey, *Control Eng. Pract.*, **1**(4), 699–714.

Åstrom, K.J. and B. Wittenmark (1989), *Adaptive Control*, Addison-Wesley, Reading, MA.

Atherton, D.P. and M. Zhuang (May 1993), Automatic tuning of optimum PID controllers, *IEE Proc. D*, **140**(3), 216–224.

Barba, D., G. Linzzo, and G. Tagliferri (1973), Mathematical model for multiflash desalting plant control, in *Proceedings of the Fourth International Symposium on Fresh Water from Sea*, Heidelberg, Germany, Vol. 1, pp. 153–168.

Bristol, E.M. (1966), On a new measure of interaction for multivariable process control, *IEEE Trans. Automat. Control*, **AC-11**, 133.

Cameron, M.C. (1996), Data reconciliation—Progress and challenges, *J. Proc. Control*, **6**(2/3), 89–98.

Corripio, A.B. (1990), *Tuning of Industrial Control Systems*, ISA, Research Triangle Park, NC.

Delene, J.G. and S.J. Ball (1971), A digital computer code for simulating large multistage flash evaporator desalting plant dynamics, Oak Ridge National Laboratory, Oak Ridge, TN, Report #ORNL-TM-2933.

El-Dessouky, H.T. and H.M. Ettouney (2002), *Fundamentals of Salt Water Desalination*, Elsevier, ISBN: 978-0-444-50810-2, 690p.

Franks, R.G.E. (1972), *Modeling and Simulation in Chemical Engineering*, Wiley, New York.

Fujii, T., O. Miyatake, T. Tanaka, T. Nakaoka, H. Matsunaga, and N. Sakaguchi (1976), Fundamental experiments on flashing phenomena in a multistage flash evaporator, *Heat Transfer Jpn. Res.*, **5**(1), 84–93.

Fukuri, A., K. Hamanaka, M. Tatsumoto, and A.S. Inohara (1985), Automatic control system of MSF process (ASCODES), *Desalination*, **55**, 77–89.

Gibilaro, L.G. and F.P. Lees (1969), The reduction of complex transfer function models using the method of moments, *Chem. Eng. Sci.*, **24**, 85–93.

Glueck, A.R. and W. Bradshaw (1970), A mathematical model for a multistage flash distillation plant, in *Proceedings of the Third International Symposium on Fresh Water from the Sea*, Dubrovnik, Yugoslavia, Vol. 1, pp. 95–108.

Gopalakrishna, S., V.M. Purushothaman, and N. Lior (1987), An experimental study of flash evaporation from liquid pools, *Desalination*, **65**, 139–151.

gPROMS (general PROcess Modelling System), 1995, Centre for Process Systems Engineering Imperial College of Science, Technology and Medicine, London, UK.

Hang, C.C., K.J. Astrom, and W.K. Ho (1991), Refinements of Ziegler-Nichol's tuning formula, *IEE Proc. D*, **138**(2), 111–118.

Hoemig, H.E. (1978), *Fichtner-Handbook on Seawater and Seawater Distillation*, Valkan-Verlag Dr. W. Classen, Essen, Germany.

Holl, P., W. Marquardt, and E.D. Gilles (1988), DIVA—A powerful tool for dynamic process simulation, *Comput. Chem. Eng.*, **12**(5), 421–426.

Husain, A., D.M.K.F. Al-Gobaisi, A. Woldai, and C. Sommariva (1993), Modelling, simulation, optimization and control of multistage flashing (MSF) desalination plants: Part I modelling and simulation, *Desalination*, **92**, 21–41, Elsevier Science Publishers B.V., Amsterdam, Netherlands.

Husain A., A. Woldai, A. Adil, A. Kesou, R. Borsani, H. Sultan, and P.B. Deshpande (1993), Modelling and simulation of a multistage flash (MSF) desalination plant, in *IDA and WRPC Conference on Desalination and Water Reuse*, Yokohama, Japan, November 3–6, 1993.

Husain, A., A. Woldai, A. Al-Radif, and R. Borsani (1993), Modeling and simulation of a multistage flash (MSF) desalination plant, in *IDA and WRPC World Conference on Desalination and Water Treatment*, Yokohama, Japan, November 3–6, 1993, Vol. III, pp. 119.

IMSL (1988), *Mathematics and Statistics Library*, IMSL Inc., Houston, TX.

Jamshidi, M. (1983), *Large Scale Systems: Modeling and Control*, North Holland, New York.

Kurdali, A., A. Moshref, A. Woldai, and D.M.K.F. Al-Gobaisi (1995), A general purpose dynamic simulation program and its application to MSF desalination plants, in *International Desalination Association IDA-95 Conference*, Abu Dhabi, United Arab Emirates.

Kurdali, A., A. Woldai, and D.M.K.F. Al-Gobaisi (1993), Knowledge based systems for desalination processes, *Desalination*, **92**, 295–307, Elsevier Science Publisher B. V., Amsterdam, Netherlands.

Lior, N. (1986), Formulas for calculating the approach to equilibrium in open channel flash evaporators for saline water, *Desalination*, **60**, 223–249.

Lopez, A.M., P.W. Murril, and C.L. Smith (November 1967), Controller tuning relationships based on integral performance criteria, *Instrum. Technol.*, **14**(11), 57.

Luyben, W.L. (1992), *Practical Distillation Control*, Van Nostrand, New York.

Marshall, S.A. (1966), An approximate method for reducing the order of a linear system, *Control*, **10**, 642–643.

Miyatake, O. and T. Hashimoto (1980), Evaporation performance of a compact multistage flash evaporator, *Kagaku Kogaku Ronbunshu*, **6**(5), 536–538.

Miyatake, O., T. Hashimoto, and N. Lior (1992), The liquid flow in multistage evaporators, *Int. J. Heat Mass Transfer*, **35**, 3245–3257.

Miyatake, O., T. Hashimoto, and C. Miyata (1983a), Analysis of liquid flow in multi-stage flash evaporators—Liquid flow pattern and pressure distribution, *Kagaku Kogaku Ronbunshu*, **9**(4), 376–382.

Miyatake, O., T. Hashimoto, and C. Miyata (1983b), Analysis of multistage flash evaporation process—Relation between liquid flow pattern and non-equilibrium, *Kagaku Kogaku Ronbun.*, **9**(4), 383–388.

Moore, B.C. (1981), Principal component analysis in linear systems: Controllability, observability and model reduction, *IEEE Trans. Automat. Control*, **AC-26**, 17–32.

Ogata, K. (2011), *Modern Control Engineering*, 5th edn., Eastern Economy Edition, PHI.

Pantelides, C.C. (1988), SPEEDUP—Recent advances in process simulation, *Comput. Chem. Eng.*, **12**(7), 745–755.

Perkins, J.D. and R.W.H. Sargent (1982), SPEEDUP: A computer program for steady-state and dynamic simulation and design of chemical process, in *Selected Topics on Computer-Aided Process Design and Analysis AIChE Symposium Series*, Vol. 78, pp. 1.

Rao, G.P. (1983), *Piecewise Constant Orthogonal Functions and Their Application to Systems and Control*, Lectures Notes in Control and Information Sciences, Springer-Verlag, Berlin, Germany.

Rao, G.P. (1993), Unity of control and identification in multistage flash desalination processes, *Desalination*, **92**, 103–124.

Rimawi, M.A., H.M. Ettouney, and G.S. Aly (1989), Transient model of multistage flash desalination, *Desalination*, **74**, 327–338.

Roehm, H.-J. (1989), *SIMFLOW—A Combined Steady State and Dynamic Simulation*, Dechema-Monographs, 116, VCH Verlagsgesellschaft, pp. 219–228.

Safanov, M.G. and R.Y. Chiang (July 1989), A Schur method for balanced-truncation model reduction, *IEEE Trans. Automat. Control*, **34**(7), 729–733.

Seborg, D.E., T.F. Edgar, and D.A. Mellichanep (1989), *Process Dynamics and Control*, John Wiley & Sons, New York.

Seider, W., D. Dowid, D. Bringle, and S. Widagdo (January 1991), Nonlinear analysis in process design, *AIChE J.*, **37**(1), 1–38.

Seul, K.W. and S.Y. Lee (1990), Numerical prediction of evaporative behaviour of horizontal stream inside a multistage flash distiller, *Desalination*, **79**, 13–15.

Seul, K.W. and S.Y. Lee (1992), Effect of liquid level on flow behaviours inside a multistage flash evaporator—A numerical prediction, *Desalination*, **85**, 161–177.

Shinsky, F.G. (1988), *Process Control System Application, Design and Tuning*, McGraw Hill, New York.

Silver, R.S. (September 1957), British Patent No. 829,819.

Smith, A.C. and A.B. Corripio (1985), *Principles and Practice of Automatic Process Control*, John Wiley & Sons, New York.

Thomas, P.J., S. Bhattacharyya, A. Patra, and G.P. Rao (1998), Steady state and dynamic simulation of multi-stage flash desalination plants: A case study, *Comput. Chem. Eng.*, **22**(10), 1515–1529.

Thye, J.F. (2010), Desalination: Can it be greenhouse gas free and cost competitive? Report of MEM Masters Project, Yale School of Forestry and Environmental Studies, New Haven, CT, May 9, 2010.

Tsypkin, Y.Z. (1978), Algorithms of optimization with a priori uncertainty (past, present, future), in *Proceedings of the Sixth IFAC Congress*, Helsinki, Finland, Pergamon Press, Oxford, UK.

Unbehauen, H. and G.P. Rao (1987), *Identification of Continuous Systems*, North Holland, Amsterdam, Netherlands.

Wilson, D.A. and R.N. Mishra (1979), Optimal reduction of multivariable systems, *Int. J. Control*, **29**(2), 267–278.

Woldai, A. (1995), Model reduction of MSF desalination plant process to first or second order with delay forms, in *ISA95, Proceedings of Automatic Control System Division*, New Orleans, LA, October 1995, pp. 343–352.

Woldai, A. (2003), *Plant Data Reconciliation, Integrated Power and Desalination Plants*, EOLSS Publishers Co. Ltd., Oxford, UK, pp. 267–282.

Woldai, A., D.M.K.F. Al-Gobaisi, R.W. Dunn, A. Kurdali, and G.P. Rao (1995a), Simulation aided design and development of an adaptive scheme with optimally tuned PID controller for large MSF seawater desalination plant (part I), in *Proceedings of the Fifth IFAC Symposium on Adaptive System in Control and Signal Processing*, Budapest, Hungary, June 14–16, 1995.

Woldai, A., D.M.K.F. Al-Gobaisi, R.W. Dunn, A. Kurdali, and G.P. Rao (1995b), Simulation aided design and development of an adaptive scheme with optimally tuned PID controller for large MSF seawater desalination plant (part II), in *Proceedings of the Fifth IFAC Symposium on Adaptive System in Control and Signal Processing*, Budapest, Hungary, June 14–16, 1995.

Woldai, A., D.M.K.F. Al-Gobaisi, R.W. Dunn, A. Kurdali, and G.P. Rao (1995c), Simulation aided design and development of an adaptive scheme with optimally tuned PID controller for large MSF seawater desalination plant (part III), in *Proceedings of the Fifth IFAC Symposium on Adaptive System in Control and Signal Processing*, Budapest, Hungary, June 14–16, 1995.

Woldai, A., D.M.K.F. Al-Gobaisi, R.W. Dunn, A. Kurdali, and G.P. Rao (1996), An adaptive scheme with an optimally tuned PID controller for a large MSF desalination plant, *Control Eng. Pract.*, **4**(5), 721–734.

Woldai, A., D.M.K.F. Al-Gobaisi, R.W. Dunn, and G.P. Rao (1995d), Simulation aided design and development of an adaptive scheme with optimally tuned PID controller for a large multistage flash seawater desalination plant, in *Proceedings of the Fourth IEEE Conference on Control Application*, Albany, New York, September 28–29, 1995.

Woldai, A., D.M.K.F. Al-Gobaisi, A.T. Johns, and G.P. Rao (1997a), ANN based adaptive control of multistage flash seawater desalination plants, in *SYSID'97 SICE*, Fukuoka, Japan, Vol. 2, pp. 907–912.

Woldai, A., D.M.K.F. Al-Gobaisi, A.T. Johns, and G.P. Rao (1997b), Optimum PID controller tuning in large MSF plants for seawater desalination, in *Proceedings of the Third Conference on Intelligent Applications in Communication and Power Systems (IACPS)*, UAE University, Al-Ain, United Arab Emirates, April 6–8, 1997, pp. 140–145.

Woldai, A., D.M.K.F. Al-Gobaisi, A.T. Johns, G.P. Rao, A. Kurdali, and A. Moshref (1995e), Process dynamics and control for MSF desalination plants—Part I (role of simulation in the design of optimal PID control), in *International Desalination Association IDA-95 Conference*, Abu Dhabi, United Arab Emirates.

Zhuang, M. and D.P. Atherton (1991), Tuning PID controllers with integral performance criteria, *Proc. IEE Conf. Control*, **1**(332), 481–486.

Ziegler, J.G. and N.B. Nichols (1942), Optimum settings for automatic controllers, *Trans. ASME*, **64**, 759–768.

Author Index

Subject Index

A

Action mode, 299
Adaptive control, 13, 35, 128
   ANNs (*see* Artificial neural
      networks (ANNs))
   definition, 299
   parameter scheduling
      adjustment mechanism, 189
      ANNs (*see* Artificial neural
        networks (ANNs))
      auxiliary variable, 188
      operating conditions, space,
        189–190
      PID controller parameters, 191
      two-dimensional space,
        operation conditions, 189–190
   single-loop PID controllers, 187
   types, 187–188
Analytical model, 30, 38, 299
Antiwindup, 299
Artificial neural networks (ANNs)
   actual and predicted outputs,
      192–193
   advantage, 193–194
   definition, 299
   integrated scheme, parameter
      scheduling, 193–194
   learning rate parameter, 198
   multilayer FNN, 195–196
   neurons, 195
   nonlinear control surface, 192–193
   optimization, 191–192
   sigmoidal function, 196–197
   two-dimensional space, operating
      conditions, 192
   weights and threshold values,
      196–197
Artificial neuron, 299

B

Backpropagation, 191, 195, 299
Ball cleaning method, 201

Biofouling, 299
Black-box approach, 30, 38, 43, 299
Blowdown, 18, 56, 76, 299
   brine level, 24–25, 64
   makeup flow control, 24
   steady-state measurements, 89
Blowthrough, 64, 97, 299
   high pressure differences, 19, 45
   hydraulic jumps, 58
   kick-plate, 58
   TBT, 125
Boiling point elevation (BPE), 54, 83,
      285, 299
Boiling point rise (BPR), 285, 299
Brackish water, 6–7, 299
Brine concentrator, 300
Brine heater, 74, 213–214, 300
   condensate level controller, 46–47
   constant parameters, equations, 48, 51
   desuperheater model, 52
   heat flux calculations, 47
   input brine temperature, 49
   liquid temperatures, steam pressure, 50
   process quantities, 48, 51
   steam flow scheme, 46
   TBT, 46
Bristol's approach, 122

C

Closed-loop control system, 19, 98, 179, 300
Computer-aided control systems
      design (CACSD), 300
Concentration, 33, 79–80, 206, 300
   black-box thermodynamic model, 43
   carbonate ions, 67
   desalting process, 7–8
   seawater makeup flow, 24
   vapor–liquid equilibrium, 73
Condensation, 201, 300
   chlorinated seawater, 17
   coefficient, tubes, 82
   distillation, 32, 53
   rate, 56

Milton Keynes UK
Ingram Content Group UK Ltd.
UKHW021631071024
449327UK00020BA/1268